DISCARD

ORANGEVILLE PUBLIC LIBRARY

THE 21ST CENTURY
The Control of Life

ALSO BY FRED WARSHOFSKY

The Rebuilt Man: *The Story of Spare-Parts Surgery*

The 21st Century: *The New Age of Exploration*

THE 21ST CENTURY **The Control of Life**

FRED WARSHOFSKY

NEW YORK  THE VIKING PRESS

Copyright © 1967, 1969 by Columbia Broadcasting System, Inc.
All rights reserved

First published in 1969 by The Viking Press, Inc.
625 Madison Avenue, New York, N. Y. 10022

Published simultaneously in Canada by
The Macmillan Company of Canada Limited

SBN 670-73583-3

Library of Congress catalog card number: 70-83253

Printed in U. S. A.

For my father . . . .

ACKNOWLEDGMENTS

Over a three-year period I was privileged to work on a unique project, *The 21st Century*, a television series that had the avowed aim of exploring the future in a way that would challenge, entertain, and inform the viewer. That the series succeeded in its aim was amply evidenced by the warm critical acclaim it received, the hundreds of letters from viewers, and the numerous awards bestowed upon it.

Unlike a book, which is very largely a personal statement and usually the work of one author, a television series is the product of a great many people: producers, writers, researchers, film editors, directors, cameramen, grips, secretaries, typists, and many others whose names never appear on the screen, but who nonetheless contribute mightily to the final product. These people were all a part of the *21st Century* television series and essential to its success. They also contributed to my own view of the future by providing and participating in a creative caldron in which the television programs were molded.

Although it is impossible to mention everyone in the *21st Century* unit, I would like to single out the following, not only for their contributions to the television series, but for the encouragement, guidance, suggestions, and research materials they provided me with for this book.

> Walter Cronkite, the narrator and principal reporter of the series
> Burton Benjamin, the executive producer
> Isaac Kleinerman, the producer
> Tom Shachtman, Judy Towers, Jon Wilkman, the researchers

There is, of course, one other group that must be mentioned in acknowledging any book on science—the scientists themselves. For these men and women are the prime movers, the activists in charting the course for the future. I have been privileged to meet some of these people in the course of my work as Science Editor of *The 21st Century*, in the writing of this book, and as a journalist covering the field of science as a reportorial beat.

In particular I should like to thank the following scientists, who graciously consented to review the individual chapters of the book for accuracy. Their review, however, in no way endorses or approves of the interpretations of scientific developments reported on herein.

> Dr. Arthur Kornberg, Chairman, Department of Biochemistry, Stanford University, Palo Alto, California
> Dr. Vincent Freda, Professor of Obstetrics and Gynecology, College of Physicians and Surgeons, Columbia University, New York, New York
> Dr. Sheldon Segal, Director, Biomedical Division, Population Council, Rockefeller University, New York, New York
> Dr. Adrian Kantrowitz, Chairman, Department of Surgery, Maimonides Medical Center, Brooklyn, New York
> Dr. Neil Miller and Dr. David Quartermain, Rockefeller University, New York, New York

CONTENTS

1.	The Creation of Life	*3*
2.	The Living Message	*19*
3.	The First Ten Months	*45*
4.	Standing Room Only	*79*
5.	Man-Made Man	*101*
6.	The Human Heart	*123*
7.	Miracle of the Mind	*145*
	Suggestions for Further Reading	*173*
	Index	*175*

THE 21ST CENTURY
The Control of Life

1 | The Creation of Life

From the time man was first aware of himself as a sentient, living being until very recently in his history, he usually considered the creation of life to be the exclusive province of the gods. Athena sprang fully formed and armed from the brow of Zeus; Marduk scattered the bloodied remains of the dragon Tiamat and in the process created not only heaven and earth, but man as well. The thundering Jehovah of the Hebrews felt life to be so important a project that He undertook the task himself, crowning the vast range of living things with man, made in His own image and possessed of "dominion over the fish of the sea, and over the fowl of the air, and over the cattle, and over all the earth, and over every creeping thing that creepeth upon the earth."

Now man has become possessed of the idea that he too can dabble in creation, that from his test tubes and technology, as well as from his loins, there might spring life. This is the quintessence of conceit, the culmination of the ancient idea of abiogenesis.

Aristotle, the synthesizer of Greek science, subscribed to abiogenesis, and out of his teachings the science of the next two thousand years was formed. The Catholic Church supported the idea of the spontaneous generation of lower forms of life. Saint Augustine said that spontaneous generation was a means by which God demonstrated his omnipotence, by interfering with the usual orderly sequence of events.

Then Church dogma was shaken by the Renaissance. Empiricism became the dominant force in science, and when a 17th-century Tuscan physician, Francesco Redi, demonstrated that the little white worms that seemed to grow spontaneously in dead meat were nothing but fly larvae, the doctrine of abiogenesis began to shake. It did not topple, however, until Louis Pasteur performed a series of brilliant experiments before the French Academy in 1864. Pasteur was determined to shoot down a theory that was known as vitalism. Developed by another Frenchman, Felix Pouchet, the theory held that some component within the air, perhaps oxygen, was the spark that could create bacteria in decaying matter. This was the vital force behind spontaneous generation, according to Pouchet.

Pasteur vehemently argued against the idea and set out to prove he was right, and at the same time to capture a prize offered by the French Academy of Sciences for a convincing experiment that would answer the question of spontaneous generation one way or another. Pasteur's approach was simple, but devastating. He maintained that the spark that created the bacteria was bacteria itself. Life, he insisted, could only arise from other life. At the Sorbonne he displayed a flask containing fermentable material. The normal open neck of the flask had been twisted into an S shape and drawn out to a point with only a narrow tip open to the air. So narrow was the opening that little dust containing microbes could enter, while the convoluted neck would further prevent any dust motes from falling into the material at the bottom of the flask. The flask had lain untouched for four years, the material inside was unfermented—proof, Pasteur insisted, that spontaneous generation was impossible.

"No," he cried, "there is no circumstance known today whereby one can affirm that microscopic beings have come into the world without germs, without parents resembling themselves. Those who claim it are the playthings of illusions, of badly done experiments tainted with errors that they did not know how to recognize or that they did not know how to avoid."

Pasteur had dealt a virtual death blow to the idea of spontaneous generation. The teaching that life arose from life became an assumption of science and dogma for the Church; few dared argue the point.

Some years before Pasteur began preparing his experiments to discredit the idea of abiogenesis, another scientist, Charles Darwin, was supporting the idea of propounding the theory of evolution, of which he gave a detailed exposition in a book called *The Origin of Species*, published in 1859. In peering down a long evolutionary tunnel, Darwin thought that at its end there simply had to be a single, primal life from which every other form of life had sprung. Such an idea inevitably led him to the question, Whence came the original species?

It was not until 1871, in a letter, that he offered a possible answer. "It is often said that all the conditions for the first production of a living organism are now present, which could ever have been present. But if (and oh! what a big if!) we could ever conceive in some warm little pond, with all sorts of ammonia and phosphoric salts, light, heat, electricity, etc., present, that a proteine compound was chemically formed ready to undergo still more complex changes, at the present day such matter would be instantly devoured or absorbed, which would not have been the case before living creatures were formed."

In fact, that warm pond is being rebuilt today in the laboratory in a determined effort to retrace the steps by which nature achieved life. Scarcely anyone doubts that by the 21st century science will have traveled that road, just as surely as nature did, and created life from a handful of "fixin's" and a lot of experimentation. For it now seems reasonably certain that life was an inevitable consequence of time and place.

If one were to attempt an approximate guess, he might estimate that life began about a billion years after the earth itself had formed. That event is subject to a great deal of conjecture, though most scientists agree that about five or six billion years ago great masses of interstellar gases and dust began to coalesce into the planets that now surround the sun. It took yet another billion or so years for the earth's mantle to coalesce. At this point, about four and a half billion years ago, the crust of the earth became stable, and it is from that event that scientists compute the age of the earth.

From then on the great physical and chemical forces interacting upon the newly formed planet all conspired to bring life into being. The great chunk of rock spinning about the sun began to generate an atmosphere of fiery gases squeezed from the solid mantle of the earth by great currents of heat produced by the radioactive decay of some elements and by volcanic eruptions that vented the earth's surface.

One of the most important theories to deal with the origin of life on earth was formulated by a Russian biologist, Alexander Oparin. In a paper published in 1924, Oparin set forth the conditions that might have existed on the primitive earth—conditions that could give rise to life. "The atmosphere at that period," he wrote, "differed materially from our present atmosphere in that it contained neither oxygen nor nitrogen gas, but was filled instead with superheated aqueous vapor. . . . The superheated aqueous vapor of the atmosphere coming in contact with the carbides reacted chemically, giving rise to the simplest organic matter, the hydrocarbons, which in turn gave rise to a great variety of derivatives."

Just what were the gases that composed Oparin's superheated aqueous vapor? What was available in that primitive atmosphere for the creation of organic molecules—the next vital step on the road to life?

One of the first scientists to attempt an experimental answer to that question was Dr. Harold Urey, a Nobel Prize-winning chemist now at the University of California. Urey took Oparin's idea of the chemical evolution of life and proposed it be tried out in the test tube. In 1953, while at the University of Chicago, Urey had a bright young chemist named Stanley Miller under his tutelage. To him, Urey suggested an experiment based upon an atmosphere he considered likely to have existed when the earth was young—methane, ammonia, water vapor, and hydrogen.

Miller constructed a sealed system of flasks and tubes into which he injected a mixture of methane, ammonia, and hydrogen. The gases were mixed with water vapor, piped in from a flask of boiling water. The entire brew was then stabbed by sixty thousand volts of high-frequency electrical sparks, a sort of man-made lightning.

After a week of this laboratory gestation, the collection flask at the end of the system was filled with water that had turned deep red. When analyzed it was found to contain several organic substances, among them amino acids.

This was proof of the theoretical first step to life the scientists thought nature had taken. For amino acids are organic molecules, the

Dr. Harold C. Urey, while at the University of Chicago in 1953, suggested that methane, ammonia, water vapor, and hydrogen were the components of the earth's primitive atmosphere.
(*University of Chicago*)

building blocks that are used in the construction of proteins. Chemists know of eighty amino acids, but only twenty are found in natural protein molecules, without question the most important structural components of all living things.

The Urey-Miller experiment produced four amino acids. Other experimenters varied the gas mixture slightly and produced other amino acids, and soon various workers were able to account for fewer than half of the amino acids common to protein. But it required a number of different attempts to get the diversity of molecules that had to be present for life to begin. And no one had successfully produced all, or even most, of the twenty amino acids in one single experimental genesis.

A great deal of controversy surrounded these atmospheric experiments, with each group certain that the gases it proposed were in fact the actual ones present at the time. All, however, included free hydrogen, the first and basic element of the universe in the primal mix. Then in 1964, at Florida State University, Dr. Sidney Fox and Dr. Kaoru Harada considered the idea of an original atmosphere that had little or no free hydrogen in it, to account for the amino acids that were very poor in hydrogen—some of the amino acids that had not been produced in all the other atmosphere experiments.

Harada put together an atmosphere of methane, ammonia, and water, with no free hydrogen to react with the intermediate products being formed from the gases on their way to becoming amino acids. He also used a different kind of energy input from those of earlier experiments. Miller had used electrical discharges, others had bombarded the gas with alpha particles and ultraviolet radiation. Harada used heat. Volcanic activity and other conditions indicated that the earth was much hotter than it is now, with thousands of hot zones well above the boiling point of water.

The result of the experiment was amino acids in profusion; twelve amino acids were formed and each was an amino acid common to protein production.

Although the controversy over the primitive atmosphere continues, all the debaters are agreed that one element was certainly *not* present then—oxygen. This lack was as important to the evolution of life as any phenomenon that did take place, for if oxygen had been present the fledgling molecules would have simply ended up as the waste products of combustion and not as the precursors of protein.

The lack of oxygen also meant that ozone, a heavy form of the oxygen, was also lacking, and the screen it now provides against most of the fierce ultraviolet rays of the sun was not present. The primordial atmosphere was bombarded every moment of daylight by ultraviolet radiation, a condition that would be lethal to living organisms, but which served as an energy source for the creation of life.

"Most of this chemistry," explains Dr. George Wald of the Biological Laboratories at Harvard University, "probably took place in the upper reaches of the atmosphere, activated mainly by ultraviolet radiation from the sun and by electric discharges. Leached out of the atmosphere over long ages into the waters of the earth, organic molecules accumulated in the seas, and there interacted with one another, so that the seas gradually acquired an increasing concentration and variety of such molecules."

About this stage of the life creation, scientists again disagree. Some hold that the simple, stately procession of time provided the opportunity for small molecules to link up to form the far more complex amino acids that are in turn the building stones of protein. Others argue that even nature must bow to some chronological pressures to speed things up.

"We have no concept as to what can happen in a million years," says Dr. Harold Urey. "We have no concept as to how large an ocean is and what an enormous amount of experimentation can take place in a large body of water over long periods of time.

"It seems to me that it may be that life originated even during the time before one could definitely say that the accumulation of the earth was complete . . . say, one hundred million years or possibly even less."

Whatever the time span, at one point or other the amino acids that leached out of the atmosphere to form what the great British biologist J. B. S. Haldane called in 1928 a "hot dilute soup" organized themselves into proteins. The process is enormously complex, for a single protein molecule is built of hundreds of amino acids alternating the basic twenty in complex and almost limitless combinations, just as the letters of the alphabet can be combined and recombined to form millions of words.

Just how these complex protein molecules were put together from the original short amino-acid chains is open to speculation. But it was an essential step in the creation of life. For only with long-chain proteins could enzymes, the catalysts that speed up chemical reactions to the point where life processes can be carried out, be constructed.

Then the self-organizing precursors of life constructed boundaries that sharply demarcated themselves from their surroundings. This has been described by Dr. J. D. Bernal of the University of London, an expert on the origin of life, as "the passage from a mere living area of metabolizing material without specific limitations into a closed organism which separates one part of the continuum from another, the living from the nonliving."

An experimental attempt to explain how such an event might have taken place was made by Dr. Sidney Fox, who now heads the Institute of Molecular Evolution at the University of Miami. Fox thinks that genetic instructions might not be necessary for protein construction. He proposes that nature could have used a chemical reaction known as thermal polymerization as the process that may have linked amino acids to form proteins. Industrial chemists have used polymerization for many years to assemble related molecules into long chains called polymers. The best example of the technology was the creation of a wholly new material, something that did not exist in nature—plastics. In a similar manner, the living organism links up amino acids, first into a small polymer called a peptide, which consists of several amino acids hooked together. The more amino acids available, the more readily polymerization occurs. The peptides are then joined to form extremely long-chain molecules that are proteins.

This process of polymerization might be compared to the action of a zipper joining together the sides of a jacket or dress. It was this zipper chemistry that Fox proposed to use in the construction of a complex protein-like compound he calls a proteinoid. "The conditions which existed to produce proteinoids are something we don't have to argue over, for they are here now."

Fox reasoned that the chemical process of polymerization was an inevitable one. The tug which pulled both halves of a zipper together would, in polymerization, be provided by heat and just as surely as the zipper came together, the amino acids would hook up to form proteinoids. Amino acids could have evolved from a primal atmosphere through a reactive process triggered by heat, by cosmic radiation, electrical jolts, or some other method, but Fox believes that the organization of these same amino acids into proteins could have occurred only by self-ordering into non-random chains.

In putting the theory to experimental test, specifically by heating amino acids, Fox violated a number of traditional chemical shibboleths. One was that if amino acids are heated above the boiling point of water, the result is only a dark unworkable mess. But he took a clue from evolutionary studies and included in his primal brew a high proportion of aspartic acid and glutamic acid, two amino acids that constitute one fourth to one half of all proteins found in

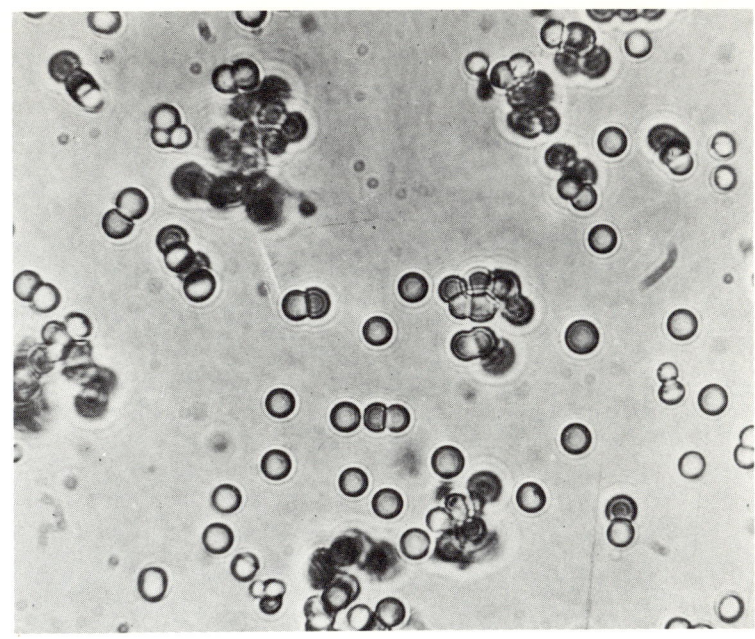

Microspheres, formed by combining in water the protein-like material called proteinoids. (*Institute of Molecular Evolution*).

nature. The solution was then heated to 150 degrees centigrade; the result was a "lightly colored material."

Chemical analysis showed the material to resemble protein molecules. An even more dramatic result occurred when these proteinoids were bathed in hot water. They took on a startling form, a spherical shape that could be clearly seen under the microscope. Moreover, there were hundreds of them in the field, bumping into each other in random motion produced by the phenomenon called Brownian movement.

Fox called these objects "microspheres." They demonstrate a long list of properties that are remarkably similar to those displayed by what he calls "contemporary cells."

"No other experiment," he wrote, "has produced anything that even begins to approach the microspheres in their similarity to cells and the properties the cells have."

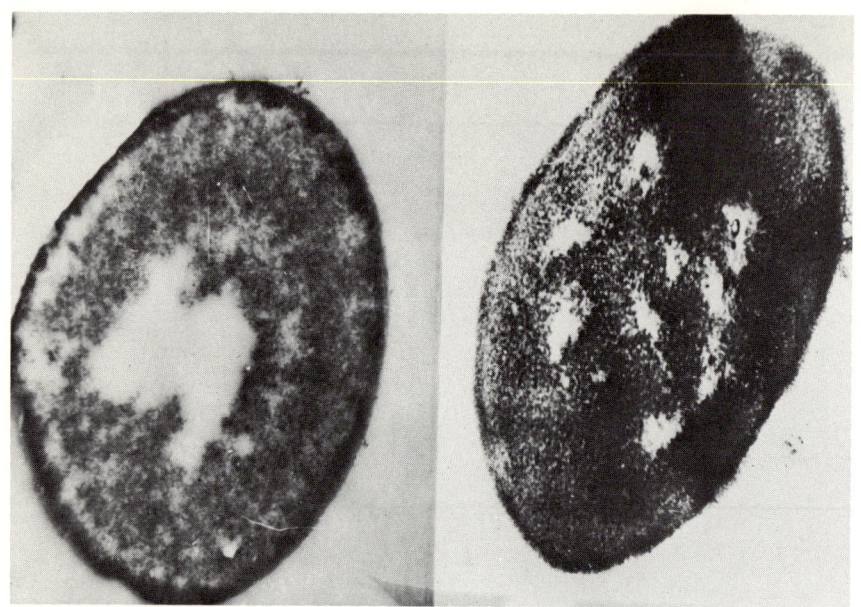

An electron micrograph of a microsphere, at right, matched with that of the bacteria *Bacillus cereus*. Experts, when presented with these micrographs, have often mistaken one for the other. (*Institute of Molecular Evolution*)

This photomicrograph shows the double boundaries that surround the microspheres. Although the walls of living cells and the microspheres' boundaries are different in a number of ways, they look remarkably similar to one another. (*Institute of Molecular Evolution*).

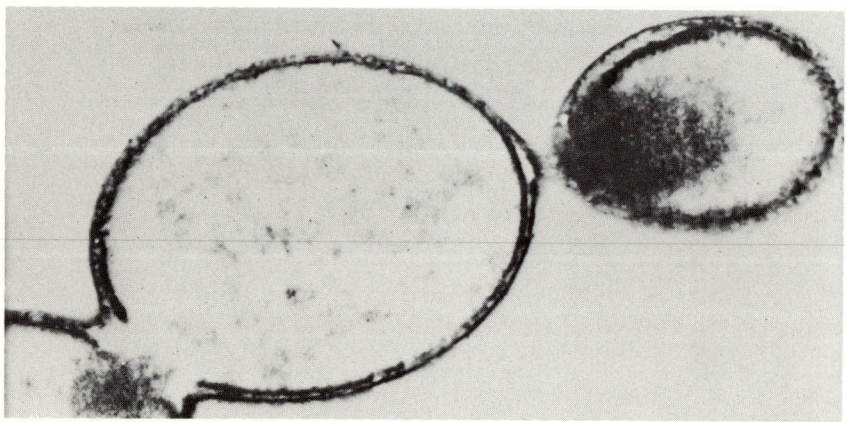

A catalogue of similarities shows microspheres to be made of the same general structural material as true cells. They are uniform in size and fall within the same size range and shape as coccoid bacteria, which some scientists believe are the most primitive bacteria. Microspheres can be treated like living cells; they can be centrifuged or sectioned for examination. They can also be stained, a classical diagnostic test for bacteria. The microspheres also have a double-layered boundary similar to the walls of living cells.

So closely do the microspheres resemble living cells, in their outward forms at least, that some of the world's best biologists have been fooled. One distinguished scientist was shown pictures of two objects taken under an electron microscope. They were both oval-shaped with double-layered walls. "Which," the scientist was asked, "is the bacterium called *Bacillus cereus*?" The scientist chose the wrong one, declaring the microsphere to be the electron portrait of the *Bacillus cereus*.

The microspheres perform a number of what, to laymen, are startlingly lifelike feats. They do not simply lie placidly in the mother liquor from which they are sprung. Although they do not possess any genetic material, as do living cells, they exhibit a form of growth and, even more remarkable, they participate in an increase of their own kind.

Many cells divide in half to reproduce, a process known as binary fission. However, some yeasts and bacteria reproduce by budding, sending out a balloon-like piece that swells and breaks off to become a new cell. Microspheres bud in a different fashion, after a week or so in the liquor. The microsphere buds do not, however, break off by themselves. They must de dislodged by external means—sending an electrical shock through the mixture, or just heating the solution. The buds are then removed from the mother liquor and placed into a new solution.

When this solution cools the proteinoid material is taken up by the buds and layered about its walls, much as an oyster deposits nacreous layers about a grain of sand. In this case, the buds are the grains of sand layering the proteinoid about themselves. After about an hour of this, the buds grow to the size of normal microspheres and begin the process again.

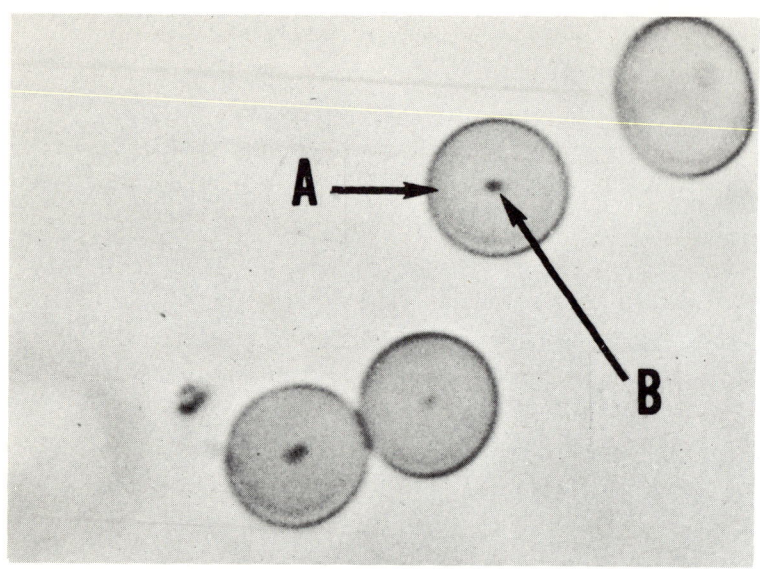

The liberated "buds" growing on their own. *A* is the proteinoid shell that has accumulated about *B*, the original "bud." (*Institute of Molecular Evolution*)

A first-generation "bud" *A*, surrounded by its proteinoid shell *B*, with a newly formed second-generation "bud" *C*. (*Institute of Molecular Evolution*)

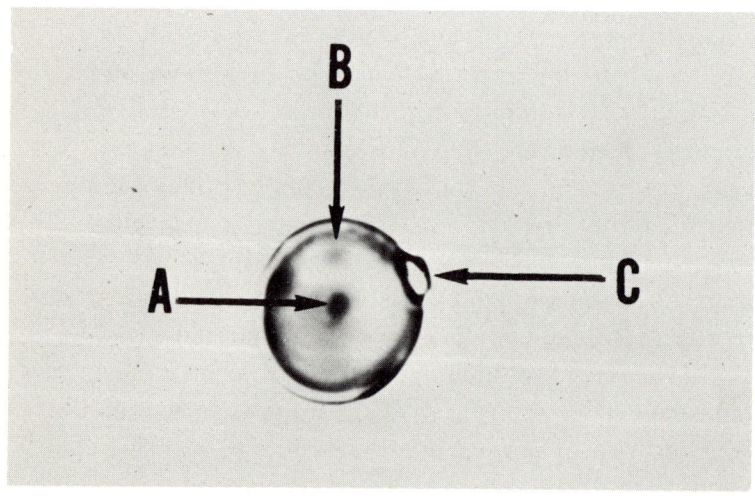

14

Although the manner of microsphere reproduction is not as complex as the budding of yeast and bacteria, it is nonetheless a replicative process.

Does this mean that the microspheres are alive? A better question might be, did they come into being before anything we dare describe as life originated? If so, and much of the evidence indicates they did, these microspheres represent an evolutionary link with the distant past, to a time before cellular forms such as we now know existed.

If the microspheres are the precursors of contemporary cells, there would still be a key step in completing the evolution from nonlife to life. A microsphere does not make its own protein within the confines of its walls. This is essentially a process governed by the nucleic acids, the DNA and RNA of the modern cell. The instructions for the assembly of protein from the amino acids taken in by the cell are carried within the DNA molecule and passed on to a portion of the cell by a messenger called RNA.

In experiments on the formation of other biologically active molecules from nonbiological materials, the vital subunits of DNA have been produced by researchers in other laboratories.

"We now seek to learn how natural experiments could have taught these organized structures to make their own protein internally with coded control by nucleic acids," explains Dr. Fox. "We believe that once this is accomplished, we will have traced the evolution of a primitive reproducing organism to a highly contemporary type of reproducing organism."

Some of the steps by which nature evolved life now seem clear and have been reproduced, in part, in the laboratory. The necessary ingredients sparked by the proper catalysts and fired by the energy of vast outpourings of ultraviolet radiation and electrical storms are organized and primed for making the enormous transition from nonlife to life.

Until very recently it was thought that this dramatic event took place only after a few billions of years of jostling and interacting among the primal amino acids.

Now scientists presume that life first emerged more than two and a half billion years ago, leaving a mere two billion years for life to evolve from the primal soup. Even this relatively short period of time is being shrunk by new discoveries. Life on earth has traditionally been dated by fossil finds, shells and bones preserved in ancient sediments and rocks. But the fossilized record of life seemed to begin about six hundred million years ago during the Cambrian period, presumably at a transition point when life forms changed from simple, fragile single cells to more organized and hardier multicelled organisms.

"There was a myth that there was nothing like a Precambrian fossil," says paleontologist J. W. Schopf. He and Dr. Elso Barghoorn of Harvard began a search for these older traces of Precambrian life. Exposing ancient rocks, some almost two billion years old, to the electron microscope, they found the faint, fossilized remains of primitive, one-celled organisms similar to the familiar protozoa. Then, in 1966, near the entrance to a South African gold mine, Dr. Barghoorn chipped a rock from the rubble turned up by a road-building crew. Under the electron microscope, Barghoorn and Schopf found evidence of a bacteria-like form of life, organized layers of spheres they called *Eobacterium isolatum* and threadlike traces they labeled "almost certainly biogenic." What made the rock and its evidence of life so important was its age—3.1 billion years old!

This ancient marker of a form of life that could thrive in an oxygen-free atmosphere appears to be a critical one for the scientists retracing the path that led from chemical evolution to biological evolution. A major new tool in that search is the discovery that minute quantities of organic compounds, a sort of residue of the original carbon-filled chemical components of the soft parts of the animal, still remain in some fossils. These so-called "chemical fossils" are not the familiar fossil structure. "The fossil molecules," explains Dr. Melvin Calvin, of the University of California, "can be extracted and identified even when the organism has completely disintegrated and the organic molecules have diffused into the surrounding material. In fact, the term 'biological marker' is now being applied to organic substances that show pronounced resistance to chemical change and whose molecular structure gives a strong indication that they could have been created in significant amounts only by biological processes."

One of the most elemental and important of these processes is photosynthesis, the process in which green plants release oxygen while reducing carbon dioxide into carbohydrates. The development of photosynthesis by living organisms would complete the transition from chemical to biological evolution and at the same time set the stage for still greater biological events to follow. For at that point a change in the gaseous make-up of the primitive atmosphere would be effected.

What were the steps that produced such a profound change? "If, as we suppose," explains Professor Wald, "life first appeared in an organic medium in the absence of oxygen, it must first have been supported by fermentations—Pasteur's 'life without air.' In so far we beg the question. But fermentation remains in a sense the basic way of life. Fermentative processes underlie all other forms of metabolism; and virtually all types of cell can survive for periods of fermentation if deprived of oxygen."

In the process of fermentation, energy is liberated and then stored by a phosphate compound within the cell called adenosine triphosphate, or ATP. At the same time, carbon dioxide is created as a waste product and released into the atmosphere, which presumably contained very little of this gas early in the earth's history. As the ages passed, the metabolic pathways of the primitive organisms became more and more sophisticated until a new process called photophosphorylation evolved, providing a means of using the sun's rays to produce ATP.

Photophosphorylation was an intermediate step, for as it evolved, the cells using it began to develop the specialized pigment called chlorophyll, which absorbs light energy, thus making the process more efficient. This set the stage for the development of photosynthesis, in which the carbon dioxide that had been merely a waste product of fermentation and other anaerobic, or oxygen-free, types of metabolism, became very important. For in the photosynthesis process, carbon dioxide is reduced to carbohydrate and oxygen is released as a by-product.

When enough oxygen had been liberated into the atmosphere by living organisms, still another great evolutionary event took place; cells learned to breathe. Respiration and photosynthesis then achieved some sort of balance, so that each could support the other. In the process, the atmosphere of the earth changed radically, with oxygen becoming a major component, twenty-one per cent, in fact, and life itself became an integral part of and contributor to the maintenance of the environment.

Soon every step in the evolution of life will have been traced, its processes described and duplicated in the laboratory. Then its creation by man becomes possible. But how shall he use this creative power? Will the 21st century see new life forms brought into being, or will man attempt to improve upon himself? Many scientists today disassociate themselves from such an idea. Many think not in terms of creating life, but simply of explaining its evolution. But full explanations will come from knowledge, and knowledge in this case will provide the power to create life.

Will man ever exercise that power? The likelihood is that he will, for never before in history has man failed to seize that which he could reach.

2 | The Living Message

The question of creating life from scratch may fade into insignificance beside the possibility of altering it to suit our own design. "We have," as one geneticist noted, "a perfectly efficient means of creating life right now, the one that nature gave us. And it's far more fun."

Altering the forms of human life may not be fun, or even a desirable endeavor, but the machinery by which it might be accomplished is being developed now. The basic transmission system, the mechanism by which our children inherit our traits—skin and hair color, disposition to disease, shape of the nose, and perhaps even intelligence—is known. We have learned to read the biochemical symbols nature uses to make up the genetic blueprints of life. By the 21st century we may be able to juggle those symbols as readily as a schoolchild assembles the letters of the alphabet to spell the words in his primer. One means of doing this is with the time-tested methods of animal husbandry, by selective breeding through many generations to preserve or eliminate certain specific traits. Eugenics, the application of similar breeding techniques in human beings, has been a popular concept for almost a century but not a working social mechanism.

"We haven't ever had any large-scale, conscious selection for better people," declares Dr. James Bonner, a biologist at the California Institute of Technology, "but we can see that of course it should be readily possible to do so. We know that several qualities of intelligence are inheritable and we know that longevity is inherited and we know that physical vigor is largely an inheritable trait, so if it were decided in a society to improve people by selective breeding, it should be possible to improve the general level of intelligence and so on and produce superpeople. The only question is how would this get started?"

Bonner foresees it as the sort of inevitable end product, a competition comparable to the space race. "Now it's clear," he adds, "that if any country on the face of the earth starts a large scale program of selecting better people, all the other countries of the world would have to get into the act too and join up, or else they'll soon be composed of obsolete, old-fashioned people like us, instead of the new superpeople, and so the idea will spread."

One of the major proponents of this idea of selective breeding was

Dr. James Bonner of the California Institute of Technology believes that it is possible to produce a nation of superpeople by the proper selection of desired genetic traits.

20

the late Dr. Hermann J. Muller of the University of Indiana. Muller was a giant among geneticists. In the 1920s he discovered that X rays could rapidly speed up the development of mutations in the geneticist's favorite laboratory animal, the fruit fly. With this tool he was able to make a number of important discoveries in basic genetics, chief among them the conviction that the X rays were producing changes in the chromosomes, the carriers of the genes, on the molecular level. For his work in genetics Dr. Muller won the Nobel Prize in 1946. He has also attracted a large following of enthusiasts who ardently support his general ideas for improving the over-all genetic health of the population, and an equally large and vocal group of opponents. From 1925 until his death in 1967 Muller made specific proposals for "improving the breed" of men. "For any group of people who have a rational attitude toward matters of reproduction," he told a symposium on human heredity at Ohio Wesleyan University a few years ago, "and who also have a genuine sense of their own responsibility to the next and subsequent generations, the means exist right now of achieving a much greater, speedier and more significant genetic improvement of the population, by the use of selection, than could be effected by the most sophisticated methods of treatment of the genetic material that might be available in the twenty-first century."

Muller's plan is to have every prospective parent deposit his genetic seed in a bank rather than to reproduce in the normal fashion. These deposits would be subject to screening, and the banks, according to Muller, would be filled with "germ cells derived from persons of very diverse types, but including as far as possible those whose lives had given evidence of outstanding gifts of mind, merits of disposition and character, or physical fitness. From these germinal stores, couples would have the privilege of selecting such material, for the engendering of children of their own families, as appeared to them to afford the greatest promise of endowing their children with the kind of hereditary constitution that came nearest to their own ideals."

Muller has included free choice in his plan, something Aldous Huxley ruled out in his *Brave New World* by eliminating parents completely. By extension then, one might imagine a young couple, newlyweds, their decision to have a baby already made, strolling down the aisle of some huge germinal supermarket. Stored on the shelves are freeze-dried packets of eggs and sperm, each neatly labeled as to chromosome content. At the end of each shelf sits a computer terminal, so that the prospective parents can call for and receive a precise accounting of the possible permutations and combinations inherent in their selection of a packet of sperm containing at least enough material for one boy with dominant genes for blond hair, blue eyes, adult height between six feet four and six feet eight, and potential intelligence quotient of 145. In small print there is a reminder that these traits are not absolutely guaranteed, but merely stand a good chance of being represented in the resultant offspring if the sperm is mated with an appropriate egg cell. The computer will help in the matchups to assure satisfaction, and Dr. Muller himself has provided a further means of producing a superior product.

"As an aid in making these choices," he wrote, "there would be as full documentation as possible provided concerning the donors of the germinal material, the lives they had led, and their relatives. The couples concerned would also have advice available from geneticists, physicians, psychologists, experts in the fields of activity of donors being considered, and other relevant specialists, as well as generalizers." Muller would also let the heat and passion that surrounded a donor's life die down before his or her seed would be placed on the shelf for selection by prospective parents. "In order to allow a better perspective to be obtained on the donors themselves, and on their genetic potentialities, as well as to minimize personality fads and to avoid risks of personal entanglements, it would be preferable for the material used to have been derived from donors who were no longer living, and to have been stored for at least twenty years."

What Dr. Muller had proposed is not antihuman, dictatorial, or unthinkable, but merely an attempt to apply to human beings the techniques of selective breeding that human beings have applied efficiently to animals for centuries. The need may become somewhat pressing by the 21st century, for we have only recently, though with enormous success, begun to circumvent nature's systematic method of selecting the genes best suited for survival in this world. Muller's idea is to put our new knowledge of human genetics to work to give nature an assist in the weeding out of decidedly undesirable genetic errors, the ones that produce people with deformed limbs and severely retarded minds and dispositions to debilitating and deadly diseases. These are the traits we now perpetuate with modern medical and surgical care, with sanitation and pesticides, with proper diets and central heating. We keep the genetic weaklings of our world alive and allow them to reproduce at will, to pass along to the next generation the same traits that nature in its inherent biological wisdom would have eliminated by killing off their owners long before they could reach the age of reproduction. We have a softer view, a shorter view of life than nature does, and our so-called compassion in this regard becomes positively myopic when we regard what biologists call the gene pool.

The gene pool is an agglomeration of hereditary traits in a constant state of flux. It is all of the possible genes residing in a total population at any one time. Thus, there are genes for, say, brown hair and harelips, for diabetes and photographic memories, for long legs and defectively short protein chains. The pool may not contain genes that carry the information likely to produce six fingers on one hand or cretinous idiots. These might be introduced as the result of genetic accidents, of mutations that alter the basic chemical information carried by the gene for a specific physical trait in individuals within the population.

The reproducibility of these traits, the mathematical rules governing their transfer from one generation to another, were first discovered by an Austrian monk named Gregor Johann Mendel, whose two hobbies were mathematics and botany. In 1857 Mendel began to crossbreed pea plants and tabulate the results. From his meticulous observations over the next eight years, he formulated a series of laws governing genetic inheritance that finally established heredity as a science, removed from the realm of speculation, mystery, and superstition. Mendel found that specific characteristics, such as height, color, and size of blossom, were passed on to the new plant by its parent. Each of the characteristics was governed by something which Mendel called "factors" and which we now call genes. These genes he found were transmitted as complete units and retained their individuality. It was a great blow to the idea then current that heredity was the result of some chance admixture or commingling of bloods of grandparents and parents united in a new combination in the offspring. Mendel, for example, found that when a tall pea was crossed with a dwarf plant,

23

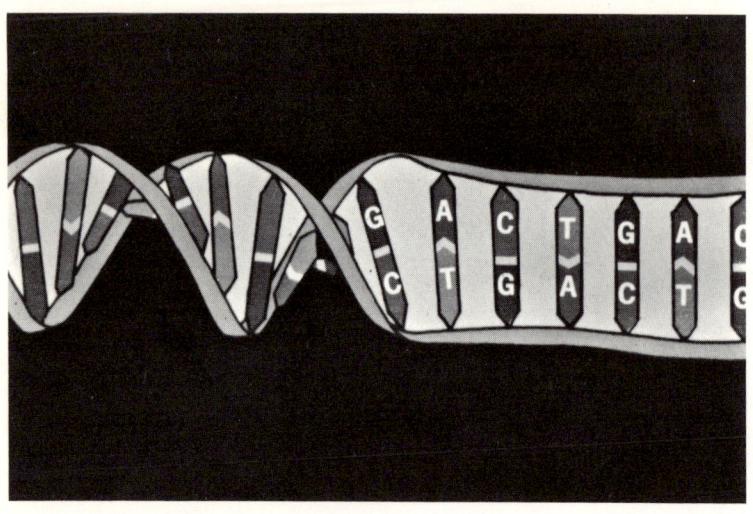

The basic carrier of genetic information is the chromosome. Coiled within the cell nucleus, like a necklace, it contains the genes, which in turn contain the specific orders that instruct each cell in what it is to become.

the progeny of the cross were either tall or dwarf, but never in-between. The haphazard-admixture theory never was demonstrated in his pea plants. But the idea stood firm for another generation, for Mendel, a modest man and an amateur scientist in his own eyes, did not have the courage to announce his results to the world.

Instead, he sent his paper, full of mathematical tables describing his results, to a famous contemporary Swiss botanist, Karl Wilhelm von Nägeli. Von Nägeli believed, as did most botanists of the time, in theories, not mathematics, and the numbers in Mendel's paper only confused him. He sent it back with a rather negative message, and Mendel, quite discouraged, mulled over the paper for a few years and then had it published by the local nature society of Brünn, the site of his monastery.

It was totally ignored, and Mendel went back to puttering in his garden, eventually became abbot of the monastery and too fat to bend over and tend his plants. He died in 1884, his work still unknown to science. Twenty years later Mendel's paper was discovered by some young biologists, and the first revolution in genetics was under way.

Now a second revolution has just taken place, perhaps the greatest in the history of science, for it has revealed the basic chemical structure of Mendel's factors, the genes, and led the way toward man's full understanding of the chemical code by which nature spells out the individual characteristics of every creature that lives. This revolution began one hundred years ago, when a German chemist named Frederick Miescher began soaking representative bits of plant and animal life in assorted acids. Each time he tried this experiment, he was left with a dark, insoluble residue, a substance that he could not identify but that he suspected of having something to do with heredity.

Optical advances over the next fifty years brought this residue into focus under the microscopes of the biologists. Within the nucleus of the cell, they found long coils, or necklaces, of dark material they called chromosomes. These, they thought, were the carriers of the hereditary material they called genes and which they likened to beads that made up the chromosomal necklace. Using the electron microscope and special X-ray techniques, scientists began to see the actual shape of the bead, which they had identified chemically as a nucleic acid called deoxyribonucleic acid, or DNA.

Experiments at the Rockefeller Institute in New York showed that DNA was indeed genetic material. In 1944 biologists at Rockefeller extracted pure DNA from a pneumococcus bacterium carrying a defective gene and transferred it to another pneumococcus. The recipient bacterium soon developed the same defect—unequivocal proof that the DNA had carried a genetic message. Over the next decade similar experiments with viruses demonstrated the same principle—genetic information was transferred by the chemical known as DNA. Every living thing, then, it seemed, from the smallest virus to the largest mammal, received its genetic instructions from DNA.

The discovery was of major importance, but it raised more questions than it answered. "Everybody wanted to know how DNA did it," said James D. Watson, one of the scientists who figured out the specific structure of the DNA molecule. In 1951, Watson was an ambitious young scientist in search of a problem to solve. "I'd just gotten my Ph.D.," he recalled for the *New York Times*, "and I knew this problem was the big one. So I got a fellowship to Copenhagen to learn about nucleic acids, since I didn't know much chemistry. I soon found out that the action was at Cavendish Laboratory in England where there was some good work in the use of X-ray crystallography."

The X-ray pictures Watson spoke of had been made not at Cavendish, but at King's College in London, by Maurice H. F. Wilkins and Rosalind Franklin. By using X rays to capture the reflections of the atoms that made up the DNA molecule, Wilkins and Franklin had come up with a series of pictures that revealed the molecule to be spiral-shaped.

Watson and the British biophysicist Francis H. C. Crick at Cavendish Laboratory proposed to build a model of the DNA molecule based upon the X-ray pictures, an approach that would perhaps spell out the mathematical laws governing DNA's operation without resorting to mathematics.

After eighteen months of model-building, Watson and Crick succeeded in constructing an elaborate structure, now world renowned as the double helix—a pair of interlacing coils linked like a spiral staircase by rungs of paired chemicals called nucleotides. For their contributions in solving this problem, Crick, Watson, and Wilkins shared the 1962 Nobel Prize for Physiology. Of their achievement, another Nobel laureate, geneticist George Beadle, wrote, "It might be said with considerable justification that the determination of the Watson-Crick structure of DNA represents the single most important development in biology of the present century."

The effects of this discovery will make possible a control over life and human affairs in the 21st century that was never dreamed of in all the previous ages of man's history. Such awesome possibilities all lie within the twisted molecule DNA, for our knowledge of its structure and the method by which it passes information on to the cell means that man will one day be able to alter the molecule at will and perhaps even to create totally new forms of life. All this in the secret of the DNA molecule.

How does it work, what is the method? DNA represents a marvel of information coding and miniaturization. If all the hereditary information present in the DNA of the total population of the earth today were collected in one place, it would form a ball no larger than the size of a pea. The secret of such incredible efficiency lies in the coding contained in the DNA molecule. The four nucleotides that make up the rungs of the DNA ladder are adenine, guanine, cytosine, and thymine: A, G, C, and T, in biochemical shorthand. The Watson-Crick model showed how these nucleotides met in pairs at the center of the helix and joined hands, holding the entire structure together and also explaining how information could be coded within the rungs. The model clearly showed that adenine could bond only with thymine and that cytosine and guanine were always linked together.

The double-helix structure has other remarkable qualities that suit it for its role as an intracellular town crier. To see how this works, one must know just a bit about cell division. The primal event that makes life possible, cell division, must occur in such a manner that the information carried by the parent cell is passed on with precision to

The DNA molecule is shaped like a spiral staircase, with railings made of a type of sugar and a phosphate, and rungs of nucleotides composed of four bases called adenine (A), thymine (T), guanine (G), and cytosine (C). The molecule is so arranged that adenine can link up only with thymine, and guanine with cytosine, but from this simple arrangement an infinite variety of coded combinations can produce the vast diversity of living things.

the daughter cells. When the cell is ready to divide, the chromosomes in the nucleus line up along a sort of spindle, a process that can be seen under a microscope using polarized light. The chromosomes stretch out as the membrane that normally surrounds the nucleus dissolves. The cell itself stretches out of shape, and at its center begins to cleave apart while the chromosomes separate and rush to each end of the dividing cell. Soon a membrane runs down the cell's center, at the point of cleavage, and the cell now splits in half, becoming two cells. Both new daughter cells possess the same information as the parent cell from which they sprang. This process is called mitosis. The hand-off of information occurred when the chromosomes split and rushed to their respective ends of the dividing cell. At that instant, the DNA within the chromosomes also split as readily as a man unzips a jacket. The chemical handclasp that held A's to T's and G's to C's was released, permitting each strand of the double helix to part and move to each end of the dividing cell. Within the daughter cell, each single strand grows a new mate, picking up raw materials from the surrounding cellular materials. Each A selects a T, each G a C, until the exact same double-stranded, twisted helical molecule of the original parent cell is reproduced in each of the daughter cells.

But it is in the arrangement of the nucleotide bases that the true elegance of DNA becomes apparent. From those four bases, the A, T, G, and C, there exists the possibility for more different combinations than there are atoms in the entire universe. In setting forth the genetic information needed by a virus, some five thousand units of A, T, G, and C may be utilized in a strand of viral DNA. In order to express the genetic heritage of one man, as many as five billion paired bases are compacted into the human DNA molecule. The problem of unraveling a code that contained this astronomical number of possibilities—a challenge to the most accomplished cryptographer—became, therefore, the next obvious goal.

Surprisingly, one of the first breaks in the code came from a nuclear physicist, the late George Gamow of the University of Colorado. Gamow proposed that the chemicals that make up the rungs of the DNA ladder be considered as playing cards. "Let's call thymines diamonds, adenines hearts, cytosines clubs, and guanines spades. If you look at the

DNA staircase now, you can see that each of the chains is just a sequence of cards which can be arranged in any order. The only rule is that if you have a heart in one chain, there has to be a diamond opposite to it in the other chain, and so on."

Gamow went on to show that if each hand dealt consisted of only three cards, the number of different suit combinations was twenty. The number twenty was the arithmetical payoff to the code, for there are twenty amino acids used by living cells to create proteins. And the ultimate goal of a DNA molecule is to provide the instructions or blueprints for the sequence in which amino acids are to be assembled in order to construct a specific protein. As the code changes, the sequence of amino acids in the protein chain is altered, so that, for example, a protein used to build brain tissue is not tailored to fit into a kidney.

The DNA code then seemed to be a triplet—each set of three base units being the code name for a specific amino acid. But the triplet code required a further refinement to make complete sense. As the three bases used to specify each amino acid had to follow a particular order along the DNA chain, there are sixty-four different ways in which the four bases can be arranged within the triplets.

This caused some confusion among biologists, physicists, and chemists all over the world who were trying to crack the genetic code. The answer was soon found, however, in nature's use of redundancy to insure reliability, as some amino acids could be spelled out by more than one triplet. The code was degenerate, a cryptographic term that described the duplication, which could only be determined experimentally. In addition, the triplet arrangement suggested by Gamow had not explained a biological fact. DNA is found only in the nucleus of the cell, but the site of protein construction, or synthesis, as the biologists call it, occurs elsewhere in the cell. The role of DNA in the assembly of amino acids into proteins might be crucial, but it was obviously not direct. DNA required an intermediary, a translator to carry its message to the site of protein assembly within the cell.

The errand boy was found to be another nucleic acid, a single-stranded form called RNA, or ribonucleic acid. It is produced when a DNA molecule unzips and makes an inverse copy of one of its strands, with a minor chemical change in its chain and in one of its bases. The thymine base is replaced in RNA by one called uracil, which also has the capability of linking up with an adenine, and its chain has a slightly different sugar from that of DNA.

The procedure then appears to be that the single-stranded RNA carries an exact message from the DNA molecule, even though one of the base letters is changed. Moreover, it carries it outside the nucleus to the site within the cell occupied by particles called ribosomes. The ribosome can be considered as a sort of automatic cryptograph. Feed in a coded tape and it will type out the message in the clear. In this case, it decodes the message carried by RNA by assembling amino acids as prescribed originally by the DNA.

With the means of transmission finally known, the next question becomes one of learning how to read the coded message. The first crack in the code came in 1961 when a young biochemist named Marshall W. Nirenberg of the National Institutes of Health took a man-made form of RNA (originally made by Severo Ochoa of New York University) that contained only one uracil triplet and fed it into a brew of several amino acids, enzymes, ribosomes, and other materials needed to synthesize proteins. The end product of this mixture was a protein made up of only one amino acid, called phenylalanine. This meant that the triplet could be entered in a decoding dictionary. A triplet consisting of three uracil bases equaled the amino acid phenylalanine.

By creating other synthetic RNA's using other triplets, additional code words were entered into the DNA dictionary. By 1966 the meaning of all sixty-four triplets had been established, including the assignment of a "nonsense" category to one triplet, which seemingly carried no message that could be translated into an amino acid. Such "nonsense" triplets were then found to be punctuation marks within the code, a sort of "stop" in a telegram, or a period at the end of a sentence, to call a halt to the protein chain.

Now, with both the medium and the message understood, the molecular biologists turned to the problem of just what to do with this marvelous new knowledge. There was, for example, an exact explanation for mutations, those usually aberrant genetic changes that produce less than desirable traits. But for centuries man has been working against nature in this regard. Everything from primitive skull openings by Stone Age healers to sophisticated surgery and drugs has worked to keep genetic misfits alive. Now there is an explanation on the molecular level of just what happens to produce a mutation or a genetic defect, and with that explanation there exists a solution, also on the molecular level. The techniques might not permit an intervention by man at this time, but in the 21st century the technology of genetic engineering may be advanced enough to repair such defects by replacing faulty parts of a DNA molecule with healthy ones.

The key to the mutation lies in a change in even a single letter of the long coded message that leads to a protein. For a protein is a very specific and absolutely essential component of life. In man there are more than one hundred thousand different types of protein. Most fall into broad classes of similar construction and purpose. One such group are the keratins, which depending upon the species, will form skin, fingernails, hair, feathers, horns, scales, even the armored shell of a turtle.

The largest and most important class of protein, however, is a group called the enzymes. These are the hurry-up agents, catalysts that speed up chemical reactions without themselves being caught up and transformed by the process. Each type of enzyme is specifically engineered for a given biochemical reaction. One group of enzymes will unlace the proteins in foods, thus freeing their amino acids for use by the cells for growth. The DNA molecule itself is serviced by its own enzymes, which aid in the reproductive process by which DNA unzips itself and replicates after cell division and in the production of RNA.

Enzymes are the proteins most needed to direct the intricate mechanisms of life. They are also the most vulnerable of the proteins to errors in coding. Just one mistake in a DNA triplet that codes only one amino acid at what is called an active site on an enzyme is enough to inhibit or prevent entirely the working function of that enzyme. Almost all mutations can be first noticed by their effects on enzymes. At least forty distinct forms of mental retardation that fall into the category of inborn errors of metabolism are the results of enzymatic failure due to a substitution of one or more letters in a DNA triplet. Phenylalanine, for example, the amino acid that played such a vital role in Nirenberg's efforts to decipher the genetic code, is essential to life but toxic to brain cells when it is not metabolized by the body. The enzyme responsible for metabolizing phenylalanine can be inhibited or completely blocked by a coding error, and the amino acid reaches toxic levels within a few weeks after birth. It then begins to destroy brain cells and produces a severe form of mental retardation called phenylketonuria.

Now a controversial test identifies victims of phenylketonuria soon after birth and a diet low in phenylalanine might prevent the damaging effects. The real hope, however, lies in tracing such errors back to their origin in the DNA or RNA molecules and correcting them there. But reaching into the nucleus of a cell to the source of life itself is a formidable undertaking.

The difficulties apparently have not scared anybody off the problem. Fifteen years ago, or just one year after Watson and Crick built their model of the DNA molecule from scraps of tin, a biochemist named Arthur Kornberg, then at Washington University in Saint Louis, took a

Dr. Arthur Kornberg, who received a Nobel Prize for his work in synthesizing DNA, believes that modified viruses might one day carry new genetic instructions into the cell and thus cure diseases caused by defective or missing genes. (*Stanford University*)

handful of chemicals off the shelf, put them into a test tube, and came up with man-made DNA. The feat, which was prodigious, earned for Dr. Kornberg the Nobel Prize in 1959, surprising most of the scientific community simply because it was awarded three years before Watson, Crick, and Wilkins got their award for constructing the DNA model. There was another surprise: Kornberg's DNA was biologically inert. Chemically and physically it was identical to the DNA found in living cells, but the test-tube variety simply lay there, passive, inactive.

It took another fourteen years to find the reason for this inactivity. Kornberg and his associates, now at Stanford University, traced the trouble to a DNA enzyme called polymerase, which zips and unzips the DNA helix, making its replication in daughter cells possible. But the polymerase used by Dr. Kornberg was contaminated with traces of other enzymes called nucleases, which produced breaks in the DNA chain. These breaks destroyed the biological activity of the test-tube DNA.

Finally, after a complex series of experiments, a pure form of polymerase was produced at Stanford.

Even with the highly purified polymerase Kornberg was unable to produce a biologically active DNA. The difficulty centered on the model being used to create it. The basic method was to break a long-chain chromosome at random, spilling DNA molecules out into solution. The polymerase would then unzip the DNA, and the separated strands would utilize the off-the-shelf chemicals Kornberg added to the mix, to construct new rails and rungs to resume the double helix form. In effect, the natural DNA served as a template for the construction of the artificial DNA. But the model Kornberg used, a bacterium called *Bacillus subtilis*, was apparently an unsatisfactory one, as its DNA molecule was too large and too complex for man-made synthesis; it becomes badly fragmented when isolated from the cell.

Kornberg finally solved the problem by using a much smaller template DNA taken from a virus. It was simpler and less subject to breakage than the bacterial DNA that the group had formerly used. Most viruses consist of a protein coat that surrounds either DNA molecules or RNA molecules. Viruses are the simplest of living organisms and may in fact be the link between life and nonlife, as they cannot reproduce except inside a susceptible host cell. A disease-causing virus

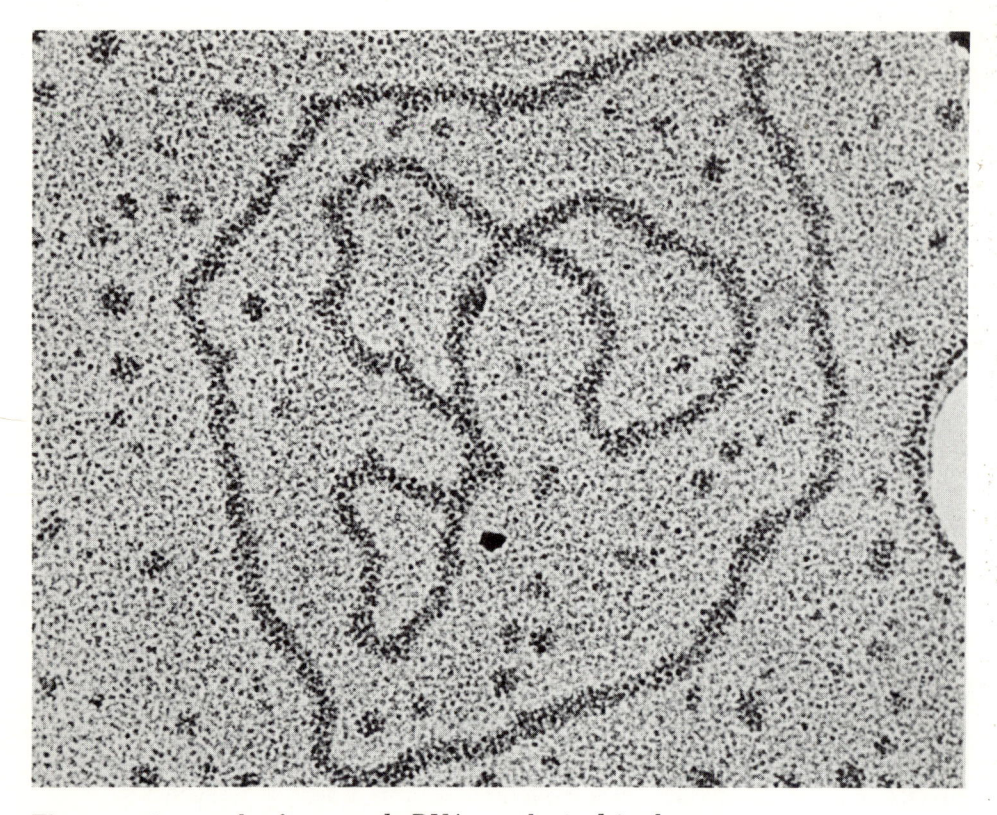

Electron micrograph of man-made DNA, synthesized in the test tube by Dr. Arthur Kornberg of Stanford University. It duplicates a natural viral DNA consisting of 5500 separate nucleotide bases. (*California Institute of Technology*)

penetrating the cell wall becomes the guiding force within the cell, supplanting the DNA and utilizing its raw materials to construct more viruses. Eventually the cell becomes so jammed with viruses it bursts and cascades millions of them onto other cells.

The virus Kornberg chose as his model is called Phi X 174. It was found by Dr. Robert Sinsheimer of Cal Tech to be a sort of dwarf whose DNA consisted, not of a double helix, but of a single strand, a simpler, more primitive form.

Phi X's role in life is to infect a bacteria called *Escherichia coli*, whose normal homestead is in the human intestinal tract and sewage.

The *Escherichia coli* is a hundred thousand times larger than Phi X and has one thousand times more DNA, but as the Bible says, "it availeth it naught." The Phi X penetrates the *Escherichia coli* and its single-stranded DNA takes over. Within twenty minutes the cell, its own genetic will replaced, has manufactured several hundred Phi-X viruses, each capable of infecting a new cell. Moreover, inside the cell, the viral DNA becomes double-stranded, like most known forms of the molecule. Using a variety of elegant experimental techniques and the DNA polymerase from the previous experiment to zip up the molecule, the Kornberg group was able to fashion a synthetic copy of the Phi-X DNA. They then separated the man-made version from the original DNA and used the synthetic DNA as a template for a second round of replication. Now this second-generation DNA was compared to the original and found to be identical with it.

The next step was to see whether it was biologically active, the stage in the experiment at which they had previously come a cropper. Even with the simple Phi-X virus, there was a large margin for error. Dr. Sinsheimer had showed that a change in just one of the Phi X's 5500 nucleotide bases would rob the virus of its infectious ability, in other words would make it biologically inactive.

Kornberg introduced his synthetic creation into a batch of *Escherichia coli*. The bacteria sickened and died, spewing forth thousands of new infectious viruses, all stemming from the original synthetic version manufactured in the test tube.

The research, according to Arthur Kornberg, represents an important step forward in understanding how viruses are duplicated when they enter cells and how DNA polymerase or similar enzymes act in the synthesis of new DNA.

"If we know how to use this enzyme to copy this particular virus then we can copy other viruses," he says, "and we can copy them in ways in which we can modify their structure by putting in alternative or fraudulent building blocks to create new forms of the virus. We can then use the synthetic virus to infect cells and produce altered responses."

These altered responses would be new genetic messages, DNA and RNA codes that differ slightly from those possessed by the cell being infected. In this way, correction of genetic errors might be made. Kornberg describes it this way:

"We can look forward to the correction of genetic defects, the cure of diseases caused by defective or missing genes. At some future date, it should be possible to cure a patient with an anemia caused by defective hemoglobin. Current treatments consist only of blood transfusions and cannot cure the disease. Assume that at some unspecified future date the gene for human hemoglobin were identified, separated from other genes and reproduced in quantity in the test tube. How could we deliver this gene into blood-forming cells of the patient? It should be possible to include this gene in one of the many viruses which infect, but do not harm us. Such a harmless virus might be exploited as a vehicle for delivering genetic information into cells where it is needed."

Cancer too might yield to such genetic engineering.

One of the possible causes of human cancer may be the virus. Although no one has yet demonstrated its presence in any form of human cancer, it seems reasonable to believe that the same mechanism that produces cancers in other mammals might also be operating in man. Hence, Kornberg has suggested experimental work with the polyoma virus, which causes a variety of malignant tumors in rats, mice, and hamsters. "On the basis of our experience," he said, "it would appear quite feasible to synthesize polyoma-virus DNA. If this synthesis is accomplished, there would seem to be many opportunities for modifying the virus DNA and thus determining where in the chromosome its tumor-producing capacity lies. With this knowledge it might prove possible to modify the virus in order to control its tumor-producing potential."

This idea of creating noninfective viruses in the test tube and then sending them in to compete with natural viruses for materials within a given cell is being pursued in a number of laboratories, even though most of the scientists working on it consider it "a long shot." At the University of Illinois one long-shot bettor is Dr. Sol Spiegelman, who successfully synthesized a biologically active RNA virus in 1965. The approach is similar to that used at Stanford by Dr. Kornberg. Instead of DNA polymerase as the zipper enzyme, however, Dr. Spiegelman uses an RNA version called replicase. The virus being copied is called *Q beta*, and for the last few years Dr. Spiegelman and his group have concentrated on stripping its RNA molecule of all instructions but one—the ability to reproduce.

This was done by constantly harvesting and regrowing in the test tube only those strands that reproduced most rapidly. After the process was repeated seventy-five times the RNA strands were shorn of such other information as how to make a viral coat of protein and were left with only the instructions for reproducing themselves. These new, synthetic RNA strands were eighty-five per cent shorter than the original, but they reproduced far more rapidly than the natural virus.

The next step was for Spiegelman to introduce his creation into bacteria already infected with the Q-beta virus. The result was a foregone conclusion. The synthetic virus, literally a reproducing machine, simply outdid the natural virus in collecting from the cell the materials needed for its own reproduction.

Such approaches, although far removed from conventional techniques for treating sick people, would fall within the framework of "doctoring." But genetic manipulation and engineering open up even more creative possibilities.

Arthur Kornberg offers this example. "Our speculations can extend even to large DNA molecules. For example, if a failure in the production of insulin were to be traced to a genetic deficit, then administration of the appropriate synthetic DNA might conceivably provide a cure for diabetes."

Kornberg's speculations presuppose an incredibly delicate refinement in technique, one we shall doubtless achieve. But even with a heavy-handed, bludgeoning approach, scientists are manipulating genes in some higher animals. At the National Institute of Genetics in Japan, Dr. Saburo Nawa performs an experiment that is being done in a number of laboratories around the world. His experimental animals are black-eyed moths. They are killed and ground up, and the DNA component of the cells that composed their bodies is chemically liberated. This DNA solution is then injected into the larva of a naturally red-eyed moth. When it hatches out it too will have red eyes, but a very small percentage of its offspring, about one in ten thousand, will have black eyes.

Nawa's technique is imprecise and inefficient, but it does work; it alters the genetic inheritance of some moths in a specific fashion. It is, without question, an example of man's control of heredity on the

37

molecular level. That control inevitably will be extended by man to his own heredity.

One enormous mystery, however, remains. No one has yet answered fully the basic question of cell differentiation—how the cell "knows" it is to be a hair cell rather than a liver or a heart cell. The problem is compounded by the fact that the DNA contained within the nucleus of every cell carries the totality of genetic information needed to produce not simply a hair or heart cell, but a complete human being. Some such system is essential to maintain biochemical order within the organism, else the forty thousand genes carried within the human chromosomes would be pouring out information at a prodigious rate, a rate that could only result in, if not ultimate chaos, then at least an incredible waste of time and chemical efficiency. What point, for example, to have a skin cell manufacture all the proteins needed for a brain cell, and vice versa?

The problem then is to find the switching mechanism that muffles some instructions and allows others to be acted upon by the cell. A partial answer comes from a pair of French scientists, François Jacob and Jacques Monod of the Pasteur Institute in Paris. Jacob and Monod have suggested that there are two kinds of genes—structural genes, which carry the coded instructions for specific proteins, and, some DNA-base lengths away, regulatory genes, which act upon the structural genes by controlling their production of RNA, the message-carrying molecule. The structural genes, like a gun, are fired by a trigger, a coded sequence at the beginning of the DNA chain that starts the gene action. This the Frenchman called the operator, and the combination of operator and structural gene they dubbed an "operon." The trigger of the operon, unlike a western gunman's weapon, was tied down by a repressor agent. The repressor prevented the operon from making messenger RNA and thus instructing the cell in protein production. The repressor is thought to be another protein called histone, which is constructed on orders from the regulator gene. It can be removed from the trigger, and the operon allowed to function by yet another molecule, an inducer that is thought to be a hormone.

What makes the system work is information feedback. Feedback is probably the simplest and most effective control system known. When more than enough of an end product is made, the excess reacts upon the starting point of the process to shut down further production. This feedback, in the form of repressor histones and inducer hormones, is thought to be the actual switching control that starts and stops the DNA tapes that turn genes on and off. This, in essence is the operon theory, and though it has not answered all the questions, it does seem to offer a working explanation of the mechanism of cell differentiation.

Once we have our hands on the switches that turn genetic information on and off, some amazing possibilities open up before us. "In the long-range future," offers the zoologist Elof Axel Carlson of the University of California at Los Angeles, "I predict the synthesis of the human genotype. The techniques for introducing this into enucleated fertilized eggs will also be developed. This will permit the necrogeneticist to bring back individuals (e.g., historical personalities) of identical genotype to the dead by using the sequences worked out from their entombed tissues. The resemblance of these individuals to ancient paintings and photographs will be startling, but their personalities will be no more like their predecessors' than are identical twins to one another."

Although Professor Carlson's prediction seems startling, even shocking, it appears likely that the techniques for doing just what he proposes will be worked out. Primitive techniques for reproducing a complex, many-celled organism from a single cell, asexually and artificially by genetic manipulation, already exist. Yet in nature, sexual reproduction involving two parents is the universal rule for most life forms more complex than one-celled organisms. Unlike mitotic division, wherein all of the genetic information is transferred intact and by separation of the DNA molecule, with both of the new daughter cells getting half of the same molecule, sexual reproduction involves a process called meiosis. Meiosis, or reduction division, is what makes sexual reproduction possible and is the process whereby nature solved the original sex problem. Parents must produce sex cells with only one half the usual number of chromosomes—forty-six in humans—so that their offspring, formed by the mating of two cells, will not wind up with twice as many chromosomes as they require. Thus, when a reproductive cell begins to divide, the twenty-three pairs of chromosomes line up exactly as they do in mitosis, but instead of dividing themselves, each chromosome pair breaks apart and seeks partners from among the other chromosomes. Then, as the cell breaks into two daughter cells, only one chromosome from each pair is carried away by the newly formed sex cell. Every sperm and egg cell thus has only half as many chromosomes as the original cell from which it has sprung, so that when later a sperm cell fertilizes an egg cell, the offspring will have a full complement of forty-six chromosomes.

It is the process of meiosis that also provides all species with their seemingly endless variations. The genetic diversity of any species, which is its greatest survival trait, is insured by the absolutely unpredictable manner in which the individual genes in each chromosome join up at conception in the formation of new gene pairs. By a completely baffling process called crossover, the genes jump from one chromosome to another and seek out new partners and begin to pour out instructions in an explosion of growth that results finally in a whole new individual. And this process is the same for frog, man, elephant, or almost any multicelled creature you care to name.

It is not, however, a process that is completely inviolate, one that cannot be altered somewhat by man's intervention. The result of this kind of genetic meddling is not the marvelous diversity that nature achieves, but an absolutely predictable and reproducible set of genetic traits that can produce carbon copies of individuals *ad nauseam.*

The process would be as follows: A human egg cell is lifted from a woman's ovary and its nucleus destroyed. A new nucleus, taken from any forty-six-chromosome cell in the body, any body for that matter, male or female, is then placed in the egg. The egg is then put in the woman's uterus, the DNA in the new nucleus activates, and the process of gestation begins. After nine months a baby will be born carrying a genetic makeup identical to that of the donor of the nucleus.

This process, called parthenogenesis, has not been attempted with humans. To date, such immaculate genetic conceptions have been limited to frogs. The guiding force in this case is Dr. J. B. Gurdon of Oxford University. Using a micropipette, Gurdon takes the nucleus from an intestinal cell and implants it within an enucleated egg cell. Something then happens to the new cell that has been created. A message seemingly flows from the egg-cell cytoplasm to the intestinal nucleus that brainwashes it free of previous associations and functions. "You are now an egg-cell nucleus and get cracking" is one possible interpretation of the message.

This frog had a mother, but no father. It was "created" by a process called parthenogenesis, in which the egg is fertilized not by a sperm, but by a nucleus taken from an intestinal cell of the mother.

About thirty per cent of these nuclear transplantations result in the production of tadpoles. Beyond that point, however, only about two per cent of the cells go on to become fully mature and fertile frogs. The problem, Dr. Gurdon feels, is in the immense difficulty of manipulating the nucleus from one cell to the other. Some subtle damage is almost always produced in the process. The difficulty in performing a similar experiment on human beings would, of course, be even greater. The frog's egg is a hundred times larger than the human's, and its nucleus lies right at the surface, not buried within the cell as is the case with most mammals. The chances of a successful parthenogenesis being accomplished in humans are, therefore, exceedingly slim, even with far more refined techniques. Gurdon has a horror of nuclear transplantation being done on human beings and resists all such speculation. His concern, he points out, is with the on-off switching of the genes and the control of genetic information within the cell. Still, the experiments have aroused intense speculation.

Many of the awesome possibilities have been summed up by, among others, John R. Platt, a biophysicist at the University of Michigan. Writing in *Science*, the journal of the American Association for the

The process of artificial parthenogenesis is extremely difficult and calls for micromanipulation of these tiny frog eggs, which are one hundred times larger than human eggs.

The leading practitioner of artificial parthenogenesis is Dr. J. B. Gurdon of Oxford University. He does not, however, see parthenogenesis as a future technique for human reproduction. His concern is strictly with the mechanisms that turn a gene on and off.

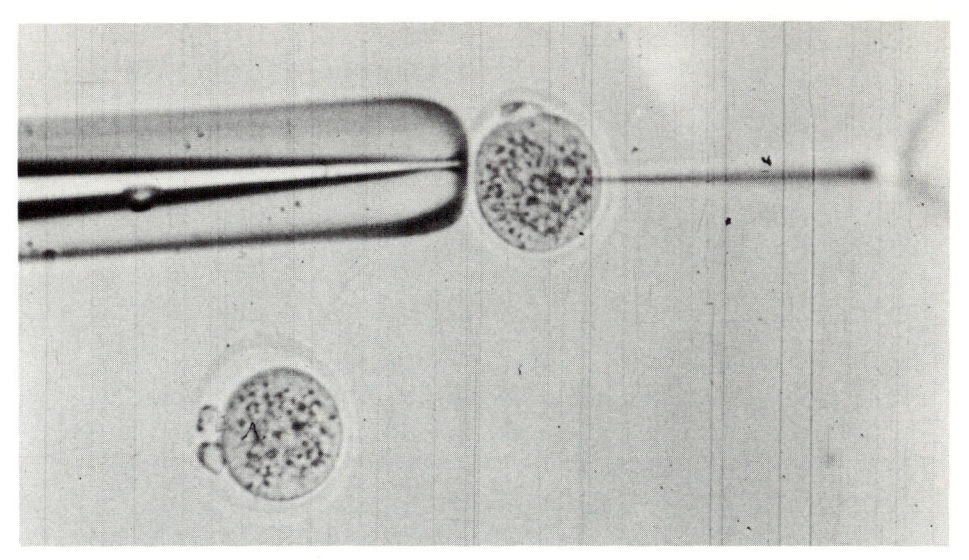

Despite the difficulties, scientists are moving ahead in developing the technology of micromanipulation. Here a rat's egg is injected with a protein substance by Dr. Teh Ping Lin of the University of California Medical Center. Dr. Lin then returns the egg to the womb and allows it to come to term. The newborn rats do not seem to have suffered from the experiments.

Advancement of Science, Platt offered the following suggestions as to these possibilities:

> For human beings, successful development of this method offers the possibility of giving babies to many couples who are unable to have children—babies which in this case could be genetic copies of the husband or wife. . . .
>
> It would also be useful to try animal-copying with the nucleus taken from one species and the egg in which it was implanted taken from another. Donkey and horse can be mated; will a donkey nucleus in a horse egg cell give a donkey—or something more like a mule? This might teach us something about the developmental embryonic differences between species. If it would work, we might be able to save some vanishing species by transplanting their cell nuclei into the egg cells of foster species. Is the DNA that carries

43

heredity destroyed immediately when an animal dies? If the meat of wooly mammoths locked for thousands of years in the Arctic ice is still edible, perhaps their DNA is still viable and might be injected, say into elephant egg cells to give baby mammoths again. By some such methods, perhaps we might achieve "paleo-reconstruction" of the ancient Mexican corn, or of "mummy wheat," or even of the flies that are sometimes found preserved in amber. One man has devoted his life to reconstructing creatures like the ancient aurochs, by backcrossing modern cattle. May not these other genetic methods of paleostudy also be worth trying? Success is uncertain, but the rewards would be great.

And uncertain too are the consequences of mishandling the tools of genetic control. In a system so carefully evolved over billions of years to pass down for generation after generation, through the millennia, the genetic information that has enabled man and every living creature now on earth to survive and evolve, what might happen when we intervene?

The geneticists and biologists working on the problem today are driven by the need to know, to understand, and to benefit man. Says Marshall Nirenberg, "When man becomes capable of instructing his own cells, he must refrain from doing so until he has sufficient wisdom to use his knowledge for the benefit of mankind."

Never before in history has man demonstrated a wisdom equal to the destructive capabilities of his technology. There seems little to indicate he has acquired that wisdom where genetic control is concerned. It may be that this will be the major accomplishment of 21st-century man, the acquisition of, if not wisdom, at least the common sense to match his technological prowess. For by the 21st century, the techniques of genetic control will almost surely have been fully developed. The question is still open.

3 | The First Ten Months

Legend holds that Julius Caesar was taken prematurely from his mother's womb by the surgical method that still bears his name. Long before that auspicious birth, however, Roman law permitted use of what is now called Caesarian section to deliver a fetus if the mother died during the last four weeks of pregnancy. Until that moment, however, when the knife slashed through the distended abdomen and muscular womb, the fetus was considered in a sort of limbo, beyond view, beyond treatment—in short, beyond all help. That idea has prevailed until almost the present moment. Since the beginnings of medicine, physicians have been taught that nature could do a far better job of protecting the fetus than could man. The fetus in the womb was an untouchable; for all practical purposes it might just as well have resided on the far side of the moon as inside the womb, separated by only a few walls of muscle from the physician's palpating fingers. But almost no doctor, from the time of Hippocrates until now, has attempted to bridge that distance.

The result, even in this the most scientific society in history, with the most sophisticated of medical technologies, has been a shockingly high level of death and defect among the four million babies born in the United States each year. Twenty-five of every thousand babies born alive will not still be alive for their first birthday. One in sixteen babies born will have a birth defect of some kind: severe mental retardation blights the lives of some 126,000 newborns each year, and, according to the American Medical Association, an "overwhelming larger number" of children are born with injuries or malfunctions of the brain that are not recognized until later, when they are slow to talk or unable to keep up with their classmates in school.

A new philosophy of pregnancy, a new daring, and an evolving technology are shattering the traditional hands-off policies of medicine, and the fetus is becoming a treatable patient, the focal point of a new medical subspecialty called "fetology."

"Fetology," explains Dr. Sidney Gellis, chief of pediatrics at Tufts University School of Medicine, "is the most exciting field in medicine today."

"The fetus," says the New Zealand obstetrician Dr. William Liley, "is just as much a patient to us as if he were lying in an incubator or crib. It just so happens that he's got his mother wrapped around him instead of a blanket."

"In the future," says Dr. Jerold Lucey of the University of Vermont, "we'll no longer assume that there is nothing we have to offer the diseased fetus. It's like outer space. The environment of the fetus is still unknown country, but the time is coming when we will know more about it. Eventually, we will probably even be able to change the environment to make it safer for the unborn child. Some risks we will have to take. Once we learn what within the uterus is a hazardous condition or a deviation from normal we may find that what we consider risky today is not risky at all."

For the fetus life is at risk almost every moment he is in the womb. Once one of the 400 million microscopic sperm fertilizes an egg, life begins. The egg, the largest cell in the body, is still at this moment a biologic mote, barely visible with the naked eye. Yet it carries from mother and father a full complement of genetic information, a pattern for growth and development that will take eighteen or more years to be fully traced.

Immediately after fertilization, the egg and sperm seem content to rest for about a day. Then cell division begins at a measured pace. The fertilized egg divides in half and yet again and again. This process,

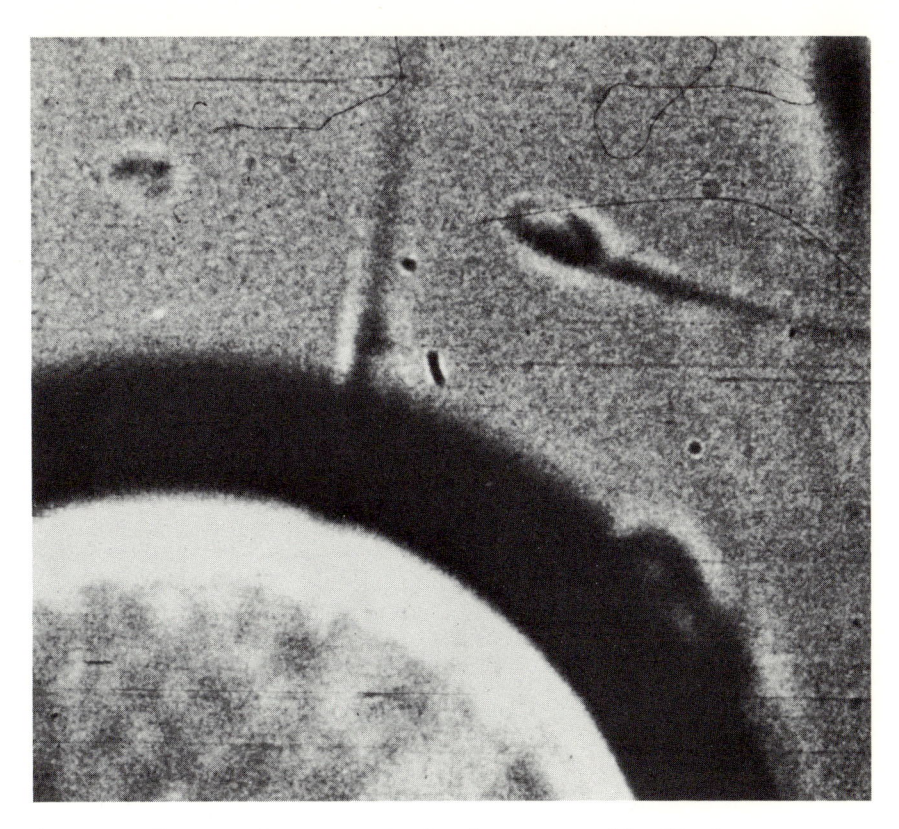

A sperm cell penetrating an egg begins the process of fertilization. This remarkable picture of the human egg and sperm is from a film made by Dr. Danielle Petrucci of Bologna, Italy.

called cleavage, will continue for about four days until the egg has been converted into a seething hollow ball composed of tiny cells.

Cleavage is a remarkable event. Unlike normal body-cell division, it does not result in an increased amount of protoplasm. The main aim of cleavage seems to be not growth but a conversion of what is essentially an outsized cell, the egg, into smaller units that are closer to the size of ordinary body cells. These smaller cells are far more maneuverable, a factor that is absolutely essential to the further development of the egg.

Although all the cells seem to be identical, a change begins to take place within them. The cells of the ball start to sort themselves out in layers. One group forms an inner layer called the endoderm. A top or outer layer called the ectoderm forms at about the same time. The middle layer of the ball, called the mesoderm, forms. Until now, the embryonic cells might have been used as components for the construction of any organ within the body, but once they are built up into layers, they are committed to specific regions, and, in fact, take on a presumptive commitment to become a stomach cell, or a brain cell, or a hair cell.

The endoderm cells, for example, begin to bud, forming outcroppings or pockets that develop into stomach, lungs, liver, intestines, in short, a primitive gut. The cells of the mesoderm assume the characteristics of blood vessels, bone, muscle, and kidneys. The ectoderm, the outer layer, folds over on itself, creating a sort of crease along the back of the embryo. The cells that make up what is now a column or tube will give rise to the spinal cord, nervous system, skin, and even the lens of the eye and what is eventually to become the crowning achievement, the brain.

The formation of these layers remains one of the great unanswered questions of the new biology. What is it that enables these cells, each containing the same blueprint encapsulated in its DNA molecules, to differentiate, to form a gut rather than a nervous system, a bone rather than a brain? One of the most elegant theories holds that the DNA molecules are arranged in series. The first series to be utilized will manufacture not only the structural proteins needed to form the cells but also enzymes that inhibit the use of the first series of DNA blueprints after they have once been used. Thus, the second series is activated and finally a third.

In this fashion a cell becomes committed. It plays out its first series option and becomes a middle-layer mesoderm cell, for example. Although it contains instructions for the other two layers as well, these were never activated and remain forever "locked" within this particular cell and all its descendants.

The increasing number of commitments grows as the cell does. The genes order increasing specialization using still other series contained within the DNA molecule and at the same time forever immobilizing the other instructions, which would follow a different route of specialization. The original commitment of the middle-layer cell is followed by all its descendants, for increasing specialization to grow into, among other things, rudimentary muscles, blood vessels, and so on.

While the middle-layer cells are increasing their specialization toward a final end product, the cells in the inner and outer layers are following similar paths toward specialization. The growth and development of the individual cells within each layer are beautifully timed so that each step in the process takes place in the proper order. Synchronization is also achieved between the different layers, for some organs are constructed from combinations of cells that originate in more than one layer.

To prove this theory of cell differentiation, the embryologists turned to some of their favorite model systems, the eggs of frogs and salamanders. One classic experiment called for the transplantation of tissue from the endoderm layer of a frog's egg to a salamander. The bit of tissue was known to give rise always to a frog's stomach if left to develop normally. In the salamander egg, the embryonic frog tissue was placed on the site where the salamander mouth normally develops. The tissue grew along with the salamander egg and eventually developed into a mouth, but true to its original genetic instructions, it was a frog's mouth, not a salamander's. Some biochemical transfer of information between the cells of the salamander and the frog tissue convinced the transplanted cells that they must become not stomach, as was originally intended, but mouth tissues. What could not be changed, however, were the species characteristics, the DNA coding that said, "You are a frog cell." Specialization then takes place at a certain time, but before that instant certain changes can be effected and a cell destined for one particular specialty can be diverted to a different path if the embryonic tracks are switched in time.

All the initial specialization of the cells begins within a matter of days, even before the egg has passed into the uterus. Everything that happens thereafter is merely an increasing amount of specialization and refinement of growth and development. It is a process that is, up to a point, common to every living creature on earth, a process that has its origins in that dim time when life moved from water to the land. The transitional creature of that exodus is the amphibian, the frog, the salamander, and others, which like the fish merely deposit their eggs and sperm in the waters that nourish and, to a degree, protect the resultant embryos.

The first creatures to make a complete break with the rivers and seas, however, were the reptiles, and they developed an ingenious method of protecting and nourishing their eggs on land. They covered them with a hard, rigid armor, a shell, that encased enough food, in the form of a yolk, for the developing embryo to live on until he was old enough to shatter the shell and assume the task of his own nourishment and protection.

The eggs of reptiles and of birds were of necessity enormous when compared to those of frogs and fish. They also had enormous consequences for the still-to-evolve mammal known as man. The egg shell, for example, and not the devil, deserves the blame for the original sexual act. "Wrapping a shell around an egg just before it is laid, as in reptiles and birds, requires previous fertilization of the egg if it is to develop," explains N. J. Berrill, Professor of Zoology at Swarthmore College. "Once the shell is formed it is too late. So in reptiles and in their descendants, the birds and mammals, the male must introduce sperm into the female reproductive tract in such a way and at such a time that sperm can reach an egg when the egg enters the oviductal tube, before it descends to where the shell is added. The happy event that we know as sex comes from this ancient requirement."

The shell forced other modifications on embryonic development. In the sea, the shell-less embryo would sprout from the surface of the yolk. Inside a shell the developing embryo grew larger and in the process became enmeshed in a series of membranes called a water jacket. Although the shell protected its inhabitant from the environment, it prevented waste products from getting out and oxygen from getting in. It was a tricky problem, for the wastes if allowed to build up would contaminate the liquid in which the embryo lived and poison it. At the same time, the rapidly growing embryo would soon use up its limited supply of oxygen, with lethal consequences. Both problems were solved by the development of a storage sac that grew out of one end of the embryo. The sac served as a dumping ground for waste materials, holding them outside the water jacket or membranes that enclosed the embryo. At the same time the sac was laced with blood vessels that carried enough oxygen to support the embryo throughout its life in the shell.

This was essentially the same design used by mammals in the formation of their eggs. The shell, however, was replaced by the body of the mother, which held the fertilized egg in protective custody and also provided both nourishment and a waste-disposal system, while the essential features such as the water jacket, yolk sac, and storage sac were retained. The yolk sac, which links us with the distant past, now serves no known function. The water jacket remains essentially unchanged, a membranous balloon encasing the embryo and the fluids that nourish it. The storage sac is the true triumph of evolutionary refinement. No longer a garbage dump, it has become a lining for the womb, a passageway for the transmission of nourishment from mother to embryo and waste from embryo to mother. The stalk that originally attached the sac to the reptilian embryo has also been retained. It becomes the umbilical cord joining the embryo and placenta together.

The placenta is perhaps the most remarkable of all the structures associated with the creation of a new being.

"Before our bodies had attained recognizable shape," declares embryologist Dr. Elizabeth M. Ramsey of the Carnegie Institution, "before our hearts had even begun to form, let alone to beat, our placentas had commenced their important business of obtaining nutrition for us from our mothers' bloodstreams."

The placenta not only performs a remarkable variety of functions but also takes over a large measure of biochemical control of the mother's body. Established in the wall of the uterus, it begins to send progesterone and estrogen, two vital female hormones, through the body. Under normal conditions, if an egg has not been fertilized, it disintegrates and the debris is sluiced out, along with the lining of the womb, as menstrual flow. The production of progesterone, normally a function of the pituitary gland, prevents that sloughing of the lining and in fact strengthens it. When a fertilized egg is implanted in the uterus, however, the placenta that forms from it takes over the production of progesterone from the pituitary and becomes in fact a foreign hand on the mother's biochemical throttle. Estrogen produced by the placenta, for example, prods the muscles of the uterus into growing longer and stronger, eventually enlarging it from the size of a large walnut to that of a small watermelon.

The placenta is also the trading post established by the fetus to deal with the mother. Across and through the vast network of blood vessels the placenta has established to nourish both the enlarged uterus and the rapidly growing embryo, pass the proteins, carbohydrates, fats, and other nutriments the embryo needs. In one sense they are stolen goods, for they are not yet completely broken down into the building blocks an individual cell can use, and so the placenta hoses them down, in some as yet unknown manner, with a spray of enzymes to the point where the embryo can digest them.

The placenta also retains its ancient reptilian function of garbage disposal, although it no longer stores the wastes it receives from the embryo. Instead, it passes them on to the mother's circulation, so that her kidneys and intestines can complete the job of waste removal.

Among the most important functions of the placenta is to act as a sort of lung, liver, and kidney for the embryo, taking oxygen from the mother's bloodstream and passing it through its own blood vessel to the creature inside its walls. It does this without ever allowing the two blood supplies—mother's and fetus's—to mix. This is accomplished by millions of tiny villi, finger-like projections that root the placenta in the womb and are the actual transfer points between mother and placenta. One means of visualizing the enormous number of villi needed to maintain the constant traffic between fetus and mother is to place them end to end. This way, they would form a carpet extending thirty miles.

In many ways, the placenta is somewhat like a spacecraft, containing a total life-support system within it and a degree of insulation and protection from the fierce environment without. The placenta houses a double-walled membrane, the amniotic sac. This contains the fluid in which the fetus floats, weightless and warm, cushioned from shock, knock, and even biological invaders. He is in effect a miniature astronaut, his one link with the outside world of his mother's body the placenta, and his link with that the umbilical cord. Through the cord passes the blood from placenta to fetus, as much as seventy-five gallons a day toward the end of pregnancy, and back again.

For all its wonders, and each is an absolutely essential service, the placenta may when fully understood help surgeons to solve one of their most baffling problems of the moment. The problem is the rejection by the body of a foreign transplant, even though the organ means life for the recipient. Somehow, the placenta circumvents the normal rejection phenomenon, and the fetus, which it nourishes and protects, is with rare exceptions free of attack from the maternal defense mechanisms. This may be because the placenta, which is not foreign to the mother's body even though it is originally a product of the embryo, maintains a rather exclusive barrier between the two, never allowing the bloods to mingle and filtering and neutralizing almost everything that passes through it.

If researchers can understand the complex subtleties by which the placenta pulls this trick off, transplant surgery may become as ubiquitous as aspirin.

Under the elaborate stewardship of the placenta, the developing embryo lacks for naught, not even, astonishingly enough, a form of medical care. Pregnant women have been smashed up in auto accidents, fallen down flights of stairs, badly injuring themselves and fracturing the limbs of their unborn babies. But the babies healed their own fractures, inside the nourishing bubble of the placenta, with no help from the outside. Fetuses have even been shot while in the womb and been subsequently born completely healed save for a scar.

Nourished and protected by the placenta, the embryo undergoes rapid changes. At four weeks of age it is barely a quarter of an inch long, but already a pair of bulges destined to expand into the forebrain can be seen. The beginnings of eyes, spinal cord, nervous system, thyroid gland, lungs, stomach, liver, kidneys, and intestines have already been laid down.

All are located in a barely recognizable head, neck, trunk, and tail formation. Budding from the trunk are two pairs of lobes that might easily be flippers, fins, or arms and legs. Along the neck runs a series of deep trenches which in the fish give rise to the gill passages and which in the human become the cranial nerves.

At this stage the most obvious organ is a rudimentary heart, beating somewhat tentatively since the eighteenth day of life. In proportion to the rest of the body, it is nine times larger than an adult heart. Surprisingly, it contains only two chambers. It is a simple double-action pump, looking and functioning exactly like that of an embryonic fish. In fact, the entire human embryo cannot be distinguished from that of any other mammal at this point in its life, a testament to the common evolutionary path traveled by all mammals. The heart, resembling that of a fish, has a powerful thrust that sends blood surging over an arterial course past the gill arches and up the neck until it reaches a thick artery that laces the spine of the embryo.

In the fifth week an internal wall begins to convert the two-chambered pump into a four-chambered heart. "The wall," explains N. J. Berrill, "forms slowly during the fifth to the eighth week of development, a three-week period that represents in terms of heart evolution something like one or two hundred million years."

During this time the embryo also begins the completion of its development from generic precursor of all mammals into a distinctly recognizable human being. At six and one-half weeks, it is more than half an inch long. Its eyes are wide and staring but without eyelids or irises. The fingers and toes look clubbed but are growing fast and have attained the first joint, but in deference to their primal ancestors remain webbed.

Internally the embryo has laid down all the organs, but they are still in various stages of development. It is still a cartilaginous, rubbery form without a bone in its tiny body.

The eighth week is the changeover point, when the embryo technically becomes a fetus. The technicality is the formation of bone cells, which begin to replace the cartilage. By the eleventh week enough bone cells have been deposited to start forming the long bones of the body—femur, radius, ulna, ribs. The walls of the trunk have grown outward from the spine like the curved ribs a ship sprouts from its keel. The ribs are joined in the front by the deck of the sternum, and the bony skeleton is virtually complete. The nervous system is meshed to the new bones and muscles, and arms and legs poke and prod the walls of the watery enclosure.

Sixteen weeks and the fetus has grown phenomenally. It measures five and a half inches and there is no doubt about its form—it is unquestionably a human baby. An equally startling change has taken place in the placenta. It no longer surrounds the double-walled amniotic sac. The rapidly growing fetus has taken over many of the placental functions for itself and has become so large as to force the placenta off to one side. From now until birth the fetus will continue to grow in size and practice the mechanical reactions it will need at birth.

The fetus becomes remarkably active, turning, stretching, and kicking in the now narrow confinement of the uterus. It learns to eat inside the womb, sucking its thumb and swallowing the amniotic fluid in which it floats. It can hear sounds and even react to them by literally jumping inside the womb when startled by a sudden, loud noise.

"The unborn is exposed to a multiplicity of sounds that range from his mother's heart beat and her voice, to outside street noises," explains Dr. Margaret Liley, director of the ante-natal clinic, National Women's Hospital, in Auckland, New Zealand, and the wife of Dr. William Liley. "Especially if his mother has not gotten too plump, a great many outside noises come through to the unborn baby quite clearly: auto crashes, sonic booms, music. And the rumblings of his mother's bowel and her intestines are constantly with him. If she should drink a glass of champagne or a bottle of beer, the sounds, to her unborn baby, would be something akin to rockets being shot off all around."

The process of growth and development proceeds from conception to birth for ten lunar months—about 280 days. As the fetus comes closer to term, and assumes more and more of a role in his own life functions, the role of the placenta is diminished correspondingly. Then it ceases to function altogether on a signal from the fetus. The signal is in the form of a hormonal message from the fetal adrenal glands. The "cease and desist" order shuts down the placenta's last remaining function, the one it began its existence with—the flow of progesterone.

The relax orders to the uterus are no longer sent, and the powerful muscles knot and unflex. The contractions eventually thrust the baby out into the world. The light blinds him, the temperature drops sharply by some twenty degrees, and his wet world grows suddenly dry. This is the world in which he must live from now on. His lungs, filled only with the salty fluid of the amniotic sac, must swiftly expel the mucus that clogs them and fill with air. There is a sudden cry, and life is resumed—this time outside the womb. A few moments later the placenta, a shriveled, tattered mass that weighs about a pound and looks somehow repulsive, is expelled from the uterine canal. It is casually tossed away, its marvelous performance, until recently scarcely recognized, even today, at the miraculous moment of birth, forgotten.

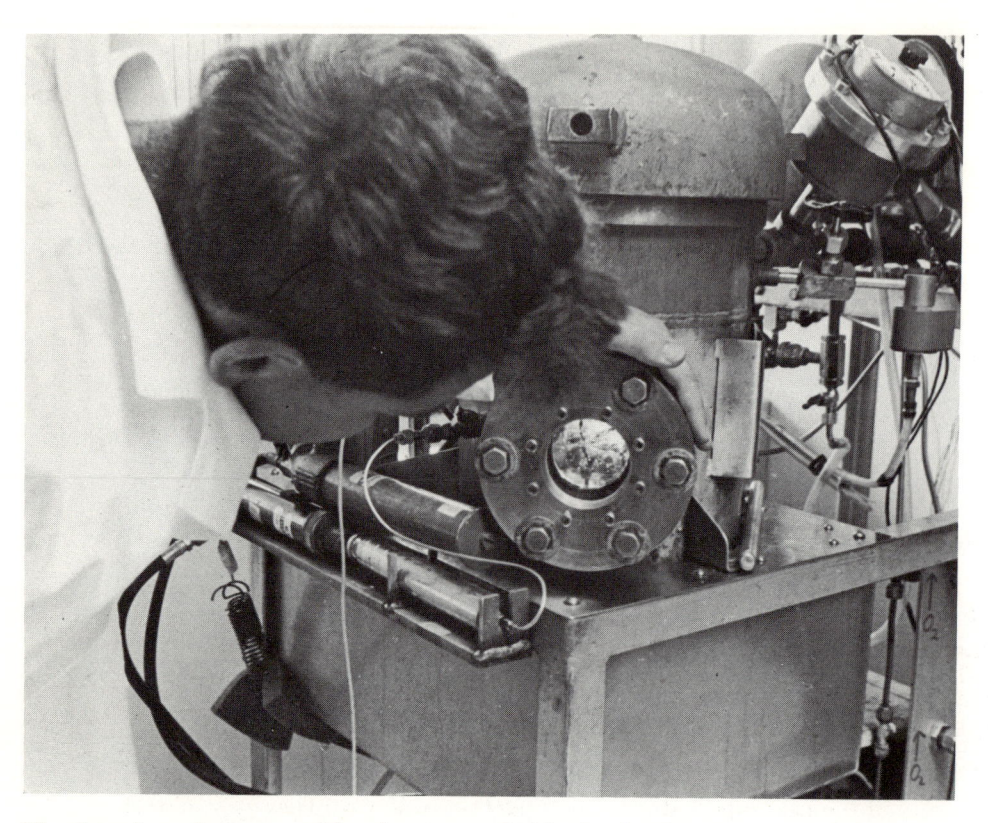

The fetus is no longer considered an untouchable. In the future, when trouble arises in the womb, the fetus may be transferred to an artificial womb. This one, developed by Dr. Robert Goodlin, of Stanford University Medical Center, is used only for research. Fetuses removed by therapeutic abortion or miscarriage are placed in the stainless-steel tank and kept alive for periods of up to forty-eight hours. (*Stanford University Medical Center*)

Despite its complexities, scientists hope to duplicate the placenta in the laboratory. If they can, it will provide a marvelous tool with which to study the fetus almost from the instant of conception to the point of delivery. What they seek is nothing less than a womb with a view.

"We should be plugging into this phenomenon of development earlier," explains Dr. Donald Pickering of the University of Nevada's laboratory of human development. "We don't at this point even know what a sick fetus looks like, but we may in the future. Once we know how a fetus becomes ill, then we will be in a position to go ahead and try to modify his environment."

Pickering plugs into the phenomenon of fetal growth with an artificial womb made of silicone rubber. His experimental subject is the monkey fetus, which he removes with the placenta and places in a silicone rubber chamber. The chamber is filled with a laboratory version of amniotic fluid, which nurtures the fetus for a day or two.

Pickering hopes to see the effect of drugs, viruses, hormones, and anesthetics on the fetus—an area of inquiry that has to date been much neglected. Tragedies such as those caused by the drug Thalidomide happen only in the absence of knowledge, and nowhere is our knowledge more deficient than during this phenomenal period of growth and development in the womb.

The Pickering womb is purely a research tool and there are no plans to use it clinically to bring aborted fetuses to term. Other artificial wombs, however, although primarily research devices, might one day be used as fetal incubators to keep the very premature alive.

At the Stanford University Medical Center, Dr. Robert Goodlin has an artificial womb that, while differing in detail, is fairly representative of how far man has gone in duplicating nature, and how far he still must go. The artificial womb, or fetal incubator, as Goodlin describes it, is built of stainless steel, about twice as big as a pressure cooker, with two viewing ports staring like eyes from the cylinder. A number of tubes feed in and out of the cylinder, pipes to carry nutrients and oxygen in and monitors to provide information on the physiological state of the fetus as it floats in synthetic amniotic fluid.

The fluid inside the chamber is under intense pressure—two hundred pounds per square inch, equal to the pressure found 450 feet beneath the surface of the sea. The pressure hammers the molecules of oxygen through the skin and into the fetal circulation—a process called cutaneous respiration, literally breathing through the skin. Most of the human fetuses Goodlin has had in his artificial womb have been young, eight, nine, and ten weeks old, the products of miscarriages. In the future such fetuses might be taken to term in such an artificial womb, but today the chambers can merely provide information and a view of fetal life for a short time. The problem lies in the incompleteness of the system. Unlike the natural placenta, which passes waste materials from the

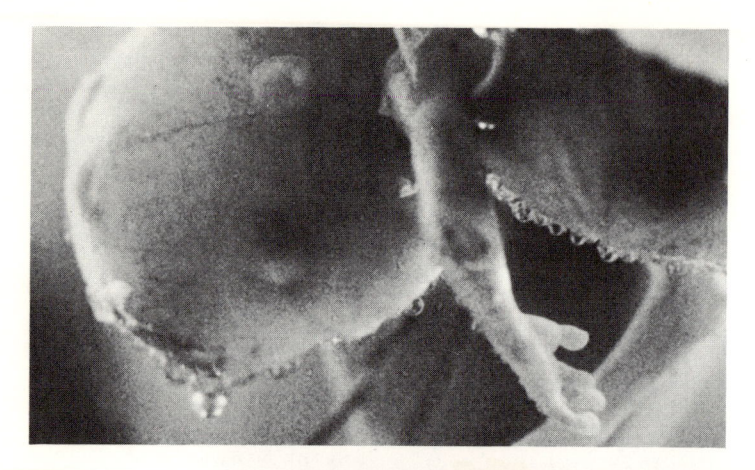

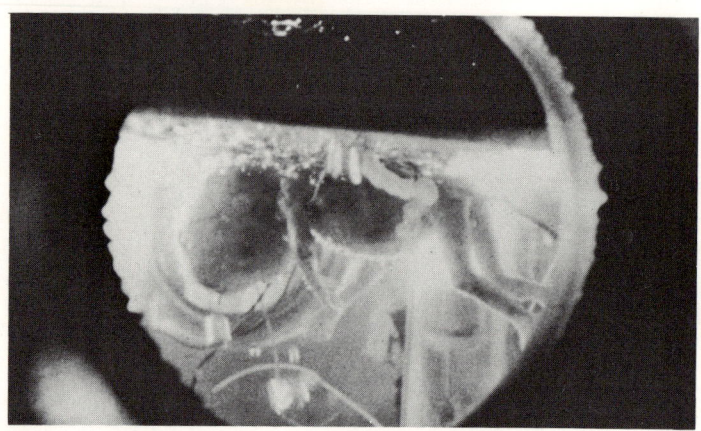

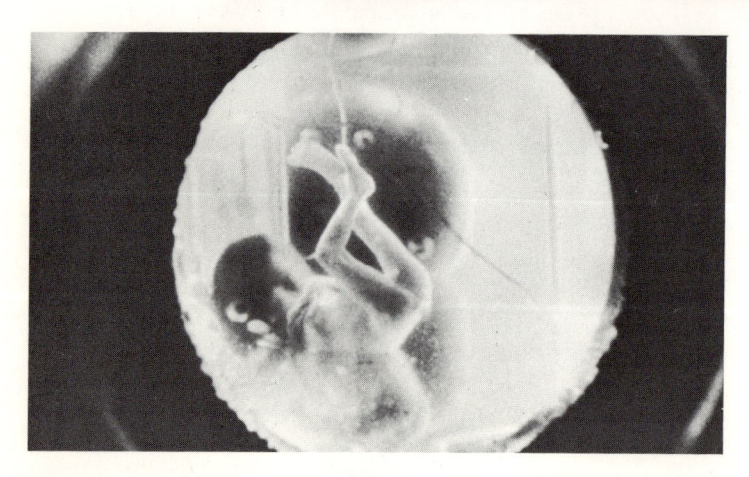

Three views of a nine-week-old fetus in Dr. Goodlin's artificial womb.

fetus to the mother for disposal, the artificial womb cannot dispose of the wastes, such as carbon dioxide, that are created by cutaneous respiration and by the normal metabolism, which is continued by the fetus outside its natural home. These problems, as yet unresolved, limit the span of life in the artificial womb to about forty-eight hours.

The point of the artificial womb is to enable the fetologist to peer in at the fetus, to watch what is happening whenever it is deemed necessary and thus be able to intervene at the proper time and save an otherwise doomed baby. X rays would provide such a view but are often considered too dangerous to the fetus.

The closer to birth the fetus is, the easier it is to view him. An intensive-care unit at the Yale–New Haven Medical Center has been equipped with an electronic monitoring system to provide the obstetrician with a continuous report on the fetal heart rate and the uterine contractions. An electrode is passed through the cervix and attached to the baby's head, while a catheter is fed past the head and into the uterus.

Compared to the old method of simply listening to the fetal heart beat by placing a stethoscope on the mother's abdominal wall, which provided only five per cent of the available information, the continuous monitoring gives the doctor ninety-nine per cent of the information available. This can alert the doctor to fetal distress, even during labor and delivery when the fetus is under the greatest stress. Of every one thousand uncomplicated pregnancies that reach the delivery room, five to seven babies die during labor.

One of the most frequent causes of fetal death and defect is due to a compression of the umbilical cord, which carries the baby's oxygen supply up to the moment it is severed from the placenta. Immediate Caesarean section used to be the treatment for babies whose oxygen supply had been pinched off owing to cord compression. Continuous fetal monitoring, however, can evaluate the effectiveness of a simple procedure—turning the mother on her side—that usually solves the problem. The monitoring provides a margin of safety so that if the fetus is still not getting the oxygen it needs, then, and only then, can the far more radical Caesarean be undertaken. The result has been a two-thirds reduction in the number of Caesareans for fetal distress at Yale.

"The results," says Dr. Edward H. Hon, chief of the section of

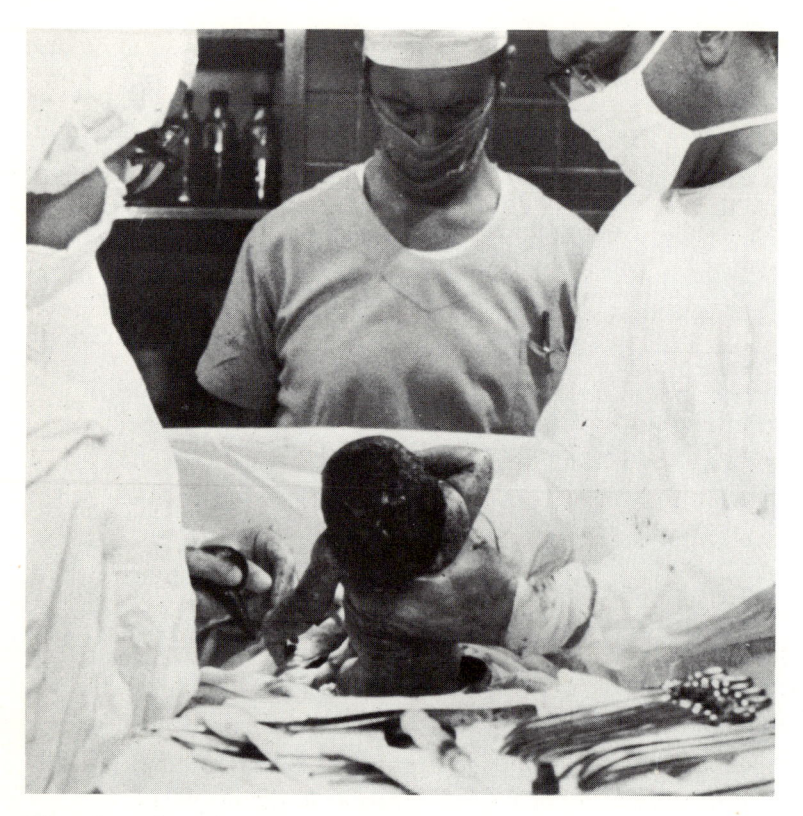

Thought to have been in use for some 2500 years, the Caesarian section still plays an important role in obstetrics.

perinatal biology at the Yale School of Medicine, "bear out what we have suspected for the past decade—that present criteria for detecting fetal distresses are untrustworthy. The introduction of an accurate method which continuously monitors the fetal heart rate and the fetal electrocardiogram provides a measure of obstetrical care that has never been achieved previously."

The burgeoning science of fetology owes its beginnings as much to a devastating genetic disease of the fetus called "erythroblastosis fetalis" or Rh disease, as to anything else. Rh disease develops only when a complex series of genetic factors is brought together at the moment of conception. First the mother must be Rh-negative, that is, her blood must be lacking a factor named after the rhesus monkey, in which it was first observed. The father must be Rh-positive, that is, he must be part of the eighty-five per cent of the human population that carries the Rh blood factor. Finally, the baby must be Rh-positive. Now, the circumstances are poised to produce Rh disease in a baby.

The disease strikes after the birth of a previous Rh-positive baby, after some of its red blood cells have spilled across the placental barrier and entered the mother's circulation. These Rh-positive cells act like a virus or bacteria and cause the body to produce antibodies that attack and destroy the foreign invaders—in this case the red blood cells of the baby. But antibody production is not an immediate reaction. First the body must be impressed with the invader as a foreign substance. Then, the next time the same invader appears, the recognition is immediate, and vast hordes of antibodies are created and attack the foreign presence. Thus, it is not the first baby that is in danger but those that come after.

Rh disease was first discovered in 1939, and with that discovery came the realization that it was killing 225,000 babies every year. The laws of probability indicated that thirteen per cent of all marriages would result in the potentially tragic union of an Rh-negative mother with an Rh-positive father.

When pregnancy occurred in such cases in those early days, obstetricians would induce premature birth in hopes of pulling the baby from the womb before it had been rendered fatally anemic. Subsequently, doctors reasoned that if the baby's blood was under attack by antibodies formed in Rh-negative blood, a valuable course of action might be to add new blood. By 1940 transfusions of Rh-negative blood following the delivery became the standard treatment of Rh disease and the death rate, which had been a staggering seventy-five per cent, dropped to about fifty per cent.

The need to wait for the baby to leave the womb meant that the conventional transfusion was of limited value. The baby's own damaged blood remained in the circulation; its red cells, shattered and half-formed, could not carry the requisite amounts of oxygen needed for survival. The toxic cycle was not broken, and doctors decided that the baby's damaged blood must be completely removed and replaced with Rh-negative blood. The technique, which was first used extensively in 1953, is called exchange transfusion, and it dramatically reduced total infant mortality from Rh disease to twenty-five per cent.

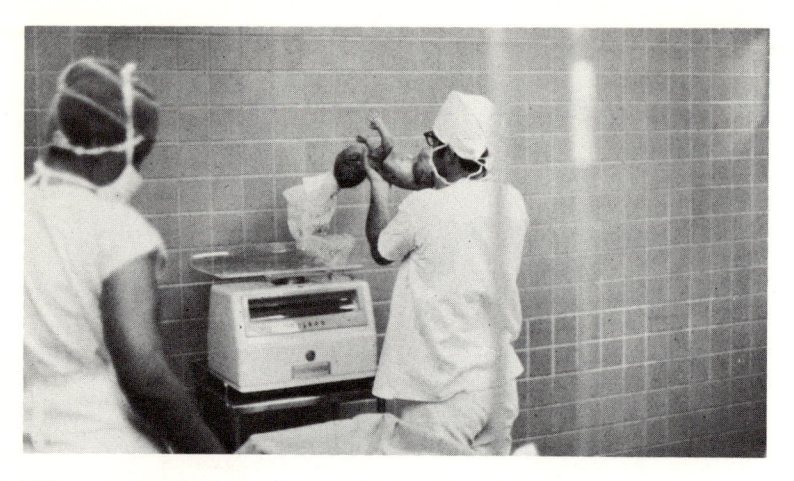

Taken surgically from the womb, a baby is carried to a scale, where its birth weight will be noted as a factor in determining medical treatment. One reason for the Caesarian section, Rh disease, may soon be eliminated by rapid and startling developments in the new field of fetology.

The exchange transfusion also helped reduce one of the dread complications of Rh disease among the children that survived. For one in ten developed severe brain damage caused by the passage of bile into the brain. When the level becomes critical it destroys brain cells. There is an automatic safeguard in the body that prevents toxic agents such as bile from being carried into the brain. It is called, appropriately enough, the "blood-brain barrier." But unfortunately the body does not appear to erect it until sometime after the first week of life. By then, for children with Rh disease and a high concentration of bile in the blood, it is too late.

A simple chemical test can determine the amount of bile in the blood, and if it is found to be above the critical level, exchange transfusions are repeated two or more times, almost like repeated rinsings to wash out a stain. The exchange transfusion made brain damage rare, but Rh disease remained, and the total mortality rate was still twenty-five per cent. Most victims died before reaching their thirty-third week of life in the womb—the minimum age before premature birth can be induced with reasonable safety. Others, brought into an environment in which prematurity robbed them of the necessary equipment for survival, did not live. Twenty-five per cent of all Rh babies died, and that seemed to be the irreducible statistic, the tragic measure of medicine's not quite successful struggle.

Research in fetology provided the next great impetus for the control of Rh disease in the late 1950s, when a British obstetrician, Dr. Douglas Bevis, found a new use for an established technique called the amniotic tap. By tapping the amniotic fluid, actually drawing a sample with a needle, chemical analysis can reveal a great deal of information about the status of the fetus. It can, for example, tell its sex, whether or not it suffers from abnormalities that would render it mentally retarded, and the severity of an Rh problem. This was a major new tool, for with it, doctors could decide exactly when to induce delivery—not unnecessarily early, forcing a prematurity crisis, and not hopelessly late, when the anemia had advanced to the point where death was inevitable.

This put the treatment of Rh disease on a more scientific basis and reduced the risk for the baby. But still, fetuses were dying in the womb, stricken by massive anemias long before they could be delivered prematurely. Amniotic taps told doctors precisely what was happening, but they were powerless to do anything. It was a frustration that could not be borne, and so Dr. William Liey broke new ground by shattering the ancient idea that the fetus was surgically inviolate. In making amniotic taps, Liley had on occasion felt his needle nick the swollen, fluid-filled abdomen of the fetus. Nothing had happened to indicate the fetus had been damaged in any way. Why not do the same thing deliberately, push the needle in another few millimeters, until it penetrated the abdominal wall and then pump in a few c.c.'s of Rh-negative blood?

Blood volume doubles every week a baby is in the womb, until at birth the baby has a total of five hundred c.c.'s of blood in its body. The new blood, pumped in during pregnancy, with its undamaged red cells to carry the desperately needed oxygen, would be immune to attack from the mother's antibodies and so enable the fetus to survive for perhaps a few weeks longer. Then the dangers of premature delivery would have been reduced enough to make the attempt. Although there was a danger to the fetus, it had to be risked—without the transfusion the tiny life was doomed.

In 1963 Dr. Liley attempted the first intra-uterine transfusion. It

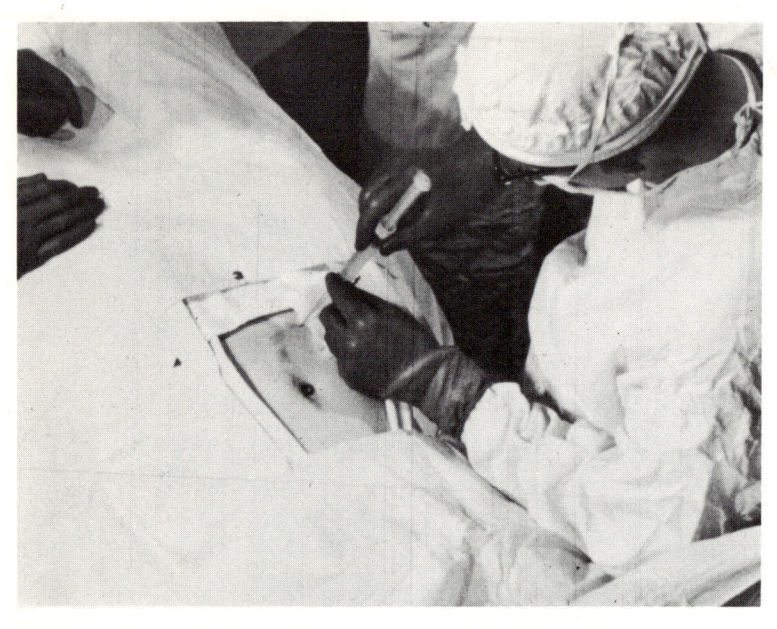

When a fetus has Rh disease, one way to provide assistance is with an intra-uterine transfusion. After X rays determine the exact position of the fetus, the mother's abdomen is punctured and the needle penetrates the fetus, to replace blood cells that have been destroyed by maternal antibodies.

failed. The fetus was so weakened by anemia, it was beyond help. The next two he tried also failed, but his fourth patient survived. An immature fetus, doomed almost from the moment of conception, survived long enough for a delivery to take place. At that point, an exchange transfusion insured his survival. That baby had been snatched from the twenty-five per cent group that dies of Rh disease.

Intra-uterine transfusion was a medical milestone, but only a handful of skilled obstetricians around the world were able or disposed to perform the delicate and dangerous procedure.

"It was a major advance," says Dr. Vincent Freda, professor of obstetrics at the Columbia Presbyterian Medical Center, "but we hoped to eliminate the disease entirely by treating the mother, not the child, before the fact, before she became burdened with antibodies."

In order to do so Dr. Freda, Dr. John Gorman, a pathologist and director of the blood bank at Columbia, and Dr. William Pollack, an immunologist at the Ortho Research Foundation, applied an idea that had been avoided and overlooked for fifty years.

"We knew," Gorman explains, "that immunity could be suppressed by giving passive antibodies." Passive antibodies are those taken from an immune person and given to someone who isn't immune. The recipient's immunological defenses are then fooled into thinking that the recipient has immunity, for there are antibodies suddenly present in the blood. But they are of a transient nature, providing only a short-term, or passive, immunity. In a few months the donated antibodies vanish from the system without imprinting upon the immunological defenses the information needed to generate an army of antibodies at the first sign of an invasion.

This principle had first been uncovered by a scientist named Theobald Smith in 1909. It was largely ignored in the years that followed, as researchers seeking new vaccines against disease sought to increase immunity rather than suppress it.

Relating the idea to Rh disease, the team of three doctors created a sort of biochemical blotter that would sweep up the red cells before they could sensitize the mother. That blotter was an antibody that would confer a passive immunity. Antibodies from women who had delivered babies with Rh disease were a ready source. Given to women immediately after they gave birth, the donated antibodies would neutralize the fetal red cells before they could challenge the mother's immunological defenses. Active, permanent immunity would be blocked.

Next came the need to answer a vital question: In what form should the antibody be given? Blood, which carries the antibodies, is a complex substance made up of a number of different components—red cells, which carry oxygen; white cells, which police the blood and keep it free of debris; platelets, which help form clots; and numerous other factors. All are suspended in an amber fluid called plasma, which makes up about fifty-five per cent of the total blood. But plasma is more than a mere carrier; it contains albumin, proteins, and, most important to the Rh problem, a complex of proteins called immunoglobulins, which carry antibodies. Often, when a temporary immunity is desired for a specific disease or problem, the entire blood plasma, with all of its antibody potential, is given. But any time plasma is injected there is the danger of also giving the recipient hepatitis. To reduce that risk, it was decided to inject only the specific fraction of the plasma, called gamma globulin, containing Rh antibodies.

Four years of research and tests on Rh-negative male volunteers at Sing Sing prison had proved the efficiency and safety of a vaccine called RhoGAM, produced from gamma globulin. Beginning in 1964, when the vaccine became available for clinical trials, more than a thousand Rh-negative women in the United States and Europe were vaccinated immediately after delivery of their first Rh-positive babies. By 1968, 159 of that group had given birth to second babies and not one of the Rh-positive babies they delivered developed Rh disease. Another group of 850 women, who served as controls, did not receive RhoGAM and gave birth to 166 babies. More than twenty per cent—thirty-five babies—in the control group suffered from Rh disease.

The idea of treating a baby even before it has been conceived is perhaps the ultimate fetology will have to offer. But other exciting possibilities are indicated by the work of Dr. Carlo Valenti, of the Downstate Medical Center in New York. Dr. Valenti is one of a handful of fetologists studying fetal chromosomes.

The procedure depends, like so many other advances in fetology, upon amniocentisis, examination of the amniotic fluid. As the fetus grows it sheds its skin gradually, depositing the cells in the amniotic fluid. Other cells are sloughed off from the bronchi, trachea, and the kidneys and bladder throughout the course of swallowing and excreting the fluid.

The fluid bearing the excreted cells is drawn off through a hollow needle. It is then placed in a centrifuge and spun until the cells all collect in the bottom of the test tube. The liquid is discarded and the cells placed in a culture medium to be grown out to the point where chromosome analyses, called karyotypes, can be made.

This elaborate and intricate procedure is called for only where preliminary examinations indicate a danger of genetic defect.

"About one in fifty babies is born with a greater or lesser degree of abnormality inherited from its parents. These weaknesses, more than five hundred of them severe enough to be classified as diseases (diabetes for example), are ordered by the genes carried in the chromosomes," explains Dr. Valenti.

A recent case illustrates the potential of karyotyping. It concerned a twenty-nine-year-old woman, sixteen weeks pregnant, with a genetic history that made her the carrier of a chromosomal abnormality that could cause mongolism in her as yet unborn child.

"A mongoloid child," explains Dr. Valenti, "—in medical terms, he is a victim of 'Down's syndrome'—has folded eyes and a flat-rooted nose (the Mongolian features from which the popular name of the anomaly derives), small head, fissured, protruding tongue, peculiarities in the lines of the palms of the hands and the soles of the feet, and retarded intellectual development ranging from idiocy to a maximum prospective mental age of seven years. A mongoloid life expectancy averages ten years—a decade of hopelessness, in most cases necessarily spent in a special institution."

Examination of the cells recovered from the amniotic fluid revealed a male fetus with a chromosomal pattern that was characteristic of mongolism. A therapeutic abortion was authorized and the autopsy and palm and sole print patterns, by which mongolism is also determined, completely supported the results of the karyotyping.

The mother fully recovered and, although saddened by the experience, is eager to have another child, provided it will be healthy. Karyotyping can give her that assurance. Decisions to interrupt a pregnancy, to perform an abortion, are difficult but obviously necessary in such cases. Unfortunately the state of fetology today is such that although we can now recognize a great number of defects and diseases of the fetus while it is still in the womb, we are almost powerless to alter them. This is one of the great future goals of this bold new science. It will be met with the aid of the geneticists, biochemists, and molecular biologists who are exploring the inner spaces of the cell.

Many genetic diseases are the result of faulty or no genetic instructions at all, instructions that are essential to the synthesis of a given enzyme. The failure of the body to make an enzyme called p-galactose-uridyl-transferase, for example, results in an inability of the newborn to digest a milk sugar called galactose. After a short time the galactose will reach toxic levels with drastic effects. Soon after birth the child has difficulty feeding and begins to vomit. Growth is retarded and death often follows from malnutrition. The babies who do survive are stunted, develop cataracts, and are severely retarded mentally, victims of an inborn error of metabolism called galactosemia.

To solve these and other genetic mistakes, the geneticists hope one day to be able to reach into the cell, into the egg or the sperm, or perhaps the just fertilized egg, and manipulate the chromosomes, to replace the faulty gene with a healthy one. The fetologist has his sights set just a bit lower. In the case of mongolism, for example, it might be possible to make up for the faulty genetic structure, already laid down in the embryo or fetus, by providing from outside the lacking enzymes. One such enzyme makes possible the construction within the body of a chemical called 5-hydroxytryptophan. It seems to have something to do with the development of normal muscle tone. Mongoloid babies are all born with flaccid muscles, and their arms and legs flap about like the limbs of a rag doll.

Recognizing this, a team of British researchers gave 5-hydroxytryptophan to fourteen mongoloid babies whose ages ranged from a few days to four months. Within a week noticeable changes had occurred, and by seven weeks normal muscle tone had been restored to thirteen of the fourteen babies. They were able to raise their heads, arms, and legs. It was obvious that something not made by the biochemical machinery of these babies had been supplied and had offset to some degree the faulty genetic instructions received from the mongoloid-inducing chromosomes.

Before anyone could leap to conclusions, the British group pointed out that there was no indication that this treatment would have any effect upon the mental development of the mongoloid children. To the fetologist, however, this is a distinct possibility, especially if the exact failure of the chromosomes involved in mongolism, or in any genetic disease, for that matter, could be delineated. Those functions could then be provided from the outside, and quite early in fetal life.

Some scientists look not only to the repair of faulty biochemical mechanisms but also to the use of drugs to improve certain capabilities of the fetus. The most obvious of these is its capacity for intellectual development. At the University of California at Los Angeles Medical Center Dr. Stephen Zamenhof has been injecting a pituitary growth hormone into the tails of pregnant rats. "We are trying to increase the number of nerve cells in the brain," he explains. "This number is fixed by the time the animal or human baby is born. Later on, only the size of the cells and their distance increases, so the brain size increases, but the number remains constant. You may go through life with the same number of nerve cells in the brain that you had when you were born."

The injection is given after the rat fetus is seven days old, or one third of the way into the gestation period. But the litters of mothers treated thus are sacrificed and their brains weighed and subjected to chemical analysis. Compared to rats born of untreated mothers, the injected group had brains that weighed thirty per cent more and possessed twenty-five per cent more cells. The density of the DNA present in the individual cells had also increased in the treated rats.

Other test groups were treated and run through a maze. In each case the rats from treated mothers performed with fewer mistakes and mastered the maze in a shorter time than did rats from a control group.

Zamenhof is the first to admit that although the results are very exciting in rats, we are a long way from trying similar experiments on humans. "This has not and probably will not be done in humans for a very long time," he says, "but there is, so far as I know, no theoretical reason why it could not be. The more complex the animal, the more pronounced the result will be."

In higher animals, experiments using other hormones have produced equally profound changes. At the Oregon Regional Primate Center in Beaverton, Oregon, Dr. Charles Phoenix and Dr. John Resko inject testosterone propinate, a male hormone, into pregnant female monkeys when the fetus is approximately thirty days old. The result when the fetus is a genetic female is quite startling. The monkey is born with the external genitalia of a male—penis, scrotal sac—and with the enlarged canine teeth that only the males have. Internally it retains an ovary, which may or may not be functional. However, chromosomal studies confirm the fact that genetically the monkey is a female. Such animals are called hermaphrodites, creatures that possess the organs of both sexes. Other researchers have achieved the same physical results, but what Resko and Phoenix have added to the experiment is a behavioral study.

"If you give the right amount of male hormone at the right time

At the Oregon Regional Primate Center, monkeys and apes are maintained in environments simulating their natural habitats. Man's closest cousins, they are the major experimental animals in fetology research. (*Don Blauvelt*)

during gestation," explains Dr. Phoenix, "then you change the brain such that when the animal is an adult it acts like a male. It acts like a male not only as far as sexual behavior is concerned, but as far as social behavior is involved. Things like aggression, for example, change. The rough and tumble play, the chasing play, the threat that is so characteristic of the male is shown by these hermaphrodites.

"Theoretically, if one interpolates our findings, they would suggest that some of the abnormalities that you find in the human may be due not simply to the external environment, whether there's a domineering mother, a submissive father in the family, but to something else, namely the hormonal levels in the mother and the fetus during development. This is what we've shown to be true for the rhesus monkey and it may very well be true for the human."

Extrapolating the Phoenix and Resko findings, it is not too difficult to foresee the day when routine screenings of the hormonal levels of all pregnant women will be undertaken. An optimum biochemical level that will create the perfect environment not only for proper sexual development but also, perhaps, for an entire range of future physical, intellectual, and behavioral characteristics, could then be maintained by the addition of drugs and other agents to mother and fetus.

The sex of children might also be not merely ascertained, or "reinforced," but determined in advance. Nature does this by assigning sex characteristics to a chromosome in both the egg and the sperm cells. The egg always carries an X or female-producing chromosome. Sperm, however, carry either an X or a Y chromosome. Should a sperm bearing an X chromosome fertilize the egg, the result will be a girl. Should one carrying a Y chromosome fertilize the egg, the product will be a boy. Once the egg and sperm have contributed their sex chromosomes, the chromosomes will appear in all future cells of this new being and thus make possible the prediction of the sex, by examination of fetal cells.

Long before the fetus forms, however, it is possible to predict sex by an examination of embryonic cells. Virtually all female body cells contain a dark spot that clings to the nuclear membrane of the cell. It is called the female sex chromatin, or Barr body, and is probably formed from one of the two X chromosomes in the cell. This is explained by the X-inactivation theory which holds that one of the two X chromosomes in every female cell is rendered inactive and coils up into a small dense mass after about the sixteenth day of gestation.

Using the Barr body as a tip-off, a pair of British researchers, Dr. Robert Edwards and Dr. Richard Gardner, of the department of physiology, Cambridge University, are selecting the sex of rabbits in advance.

Rabbits ordinarily pop eggs from the ovaries during copulation. The egg enters the oviduct, the rabbit equivalent of the human Fallopian tube, and begins its journey to the womb. Fertilization of the egg takes place within the oviduct by sperm that swim up the vaginal canal and into the tube. The fertilized egg immediately begins to divide in half, again and again, until it consists of several thousand cells. At this point it is called a blastocyst; its sex is now determined, and it is ready to begin the next step in embryonic development—implantation in the womb.

At this point Edwards and Gardner intervene, inserting a pipette, a slender glass tube, into the womb and sucking out the fertilized egg. In the case of the rabbit, there are usually six or eight eggs, a major factor in choosing it as the experimental model. From these eggs the researchers then extract with extreme delicacy a hundred or so of the microscopic cells, as yet undifferentiated, that make up the blastocyst. They search them for Barr bodies and thus determine the sex of the embryos.

Once the sex has been determined, it is a simple matter to throw away that which is unwanted, and restore to the mother's uterus the embryos of the desired sex.

The procedure is of course far more difficult to perform than is the description. Of eighty embryos in one series, forty-eight failed to implant in the uterine lining after being returned to the womb, and fourteen died after implantation. But eighteen went to full term and bore out the original sex prediction.

Even when the technique is refined to the point where all the embryos can be safely restored to the womb and come to term, it is unlikely to be applied to humans. For one thing, the human female usually releases but one egg a month. Fertility drugs, which induce superovulation and the production of four, five, and even six eggs at a time, could conceivably get around this barrier, but there are other problems. The Barr body does not seem to appear before the human embryo has implanted itself in the womb. Karyotyping, the means used to determine the sex of fetal cells, requires far more cells than are available in the human blastocyst just before implantation. Gardner and Edwards suggest a way out of this impasse: snatching the egg early in its cleavage stages and dividing it in half. One half would then go on to develop normally and the other, which would also be nurtured and encouraged to continue its growth, could be used for analysis.

Why go through such elaborate and highly dangerous procedures to select one sex or another? Why relate abortion to sex? Explain Edwards and Gardner: "When it becomes possible to determine the sex of human blastocysts with certainty, and also to identify quite certainly women who are heterozygous (carriers) for a sex-linked disease, then it will be a simple matter to ensure that no such woman bears a male child. Many of the genetic diseases such as haemophilia, a form of muscular dystrophy, and many enzyme-deficiency diseases are sex-linked and occur almost always in boys. Would not the nonreplacement of a blastocyst be (socially and ethically) far more acceptable than a full-scale abortion of the implanted, thriving fetus—the only alternative available today?"

These and other questions never before raised will cause agonies of moral and ethical wrangling as the technology of fetology grows. Yet within the technology there might be some means of resolving such problems even before they arise. What if one could choose sex in advance of fertilization? Since sex is for all practical purposes determined by the chromosomes carried in the male sperm, why not examine the sperm to see if some revealing feature might disclose its chromosomal nature? This is precisely what Dr. Landrum B. Shettles of Columbia University's College of Physicians and Surgeons asked. The answer he found after a series of remarkable experiments was that the X-bearing sperm have large, oval heads and move more slowly than do the Y-bearing sperm, which are smaller and have round heads.

Shettles' work goes a long way toward explaining why approximately 106 boys are born to every 100 girls. Of the approximately 400 million sperm that compose each ejaculation, about sixty-six per cent generally carry the Y chromosome, and since sperm containing Y chromosomes travel faster than those containing X chromosomes, the likelihood of a boy resulting from fertilization is slightly increased. There is still another factor that accounts for the ratio of boys to girls. Dr. Shettles studied family trees in which male offspring predominate and then analyzed semen samples from members of these families. He found an unusually high percentage of Y-bearing sperm, as many as ninety-six per cent in some cases. Where girls predominate, Dr. Shettles found the X-carrying oval-headed sperm in the vast majority.

Taking the two basic differences, shape and speed, into account, it might one day be possible to develop a filter, perhaps somewhat like an intra-uterine contraceptive device, that would screen out either male- or female-carrying sperm—depending on which was desired. Or the sperm might be separated in the test tube and only one type artificially introduced into the vaginal canal—artificial insemination. Or egg and sperm might be mated in the test tube and subsequently implanted into the womb.

Choice of sex might act as a control on family size, offering parents a boy and a girl and removing the need to "keep trying" for a boy or girl to complete the family.

None of the possibilities for embryonic manipulation is considered improbable, and a number of them have already been demonstrated in

Pregnant monkeys such as this one receive outstanding care and are subjected to a variety of experiments at the Oregon Primate Center. The aim is to learn as much as possible about techniques and environmental conditions that will lead to improvements in the care and treatment of human fetuses. (*Don Blauvelt*)

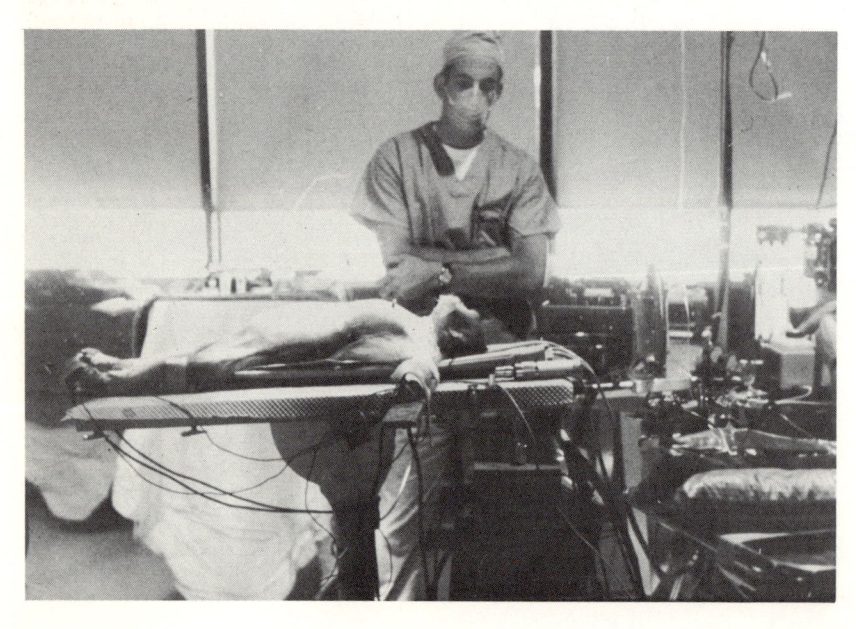

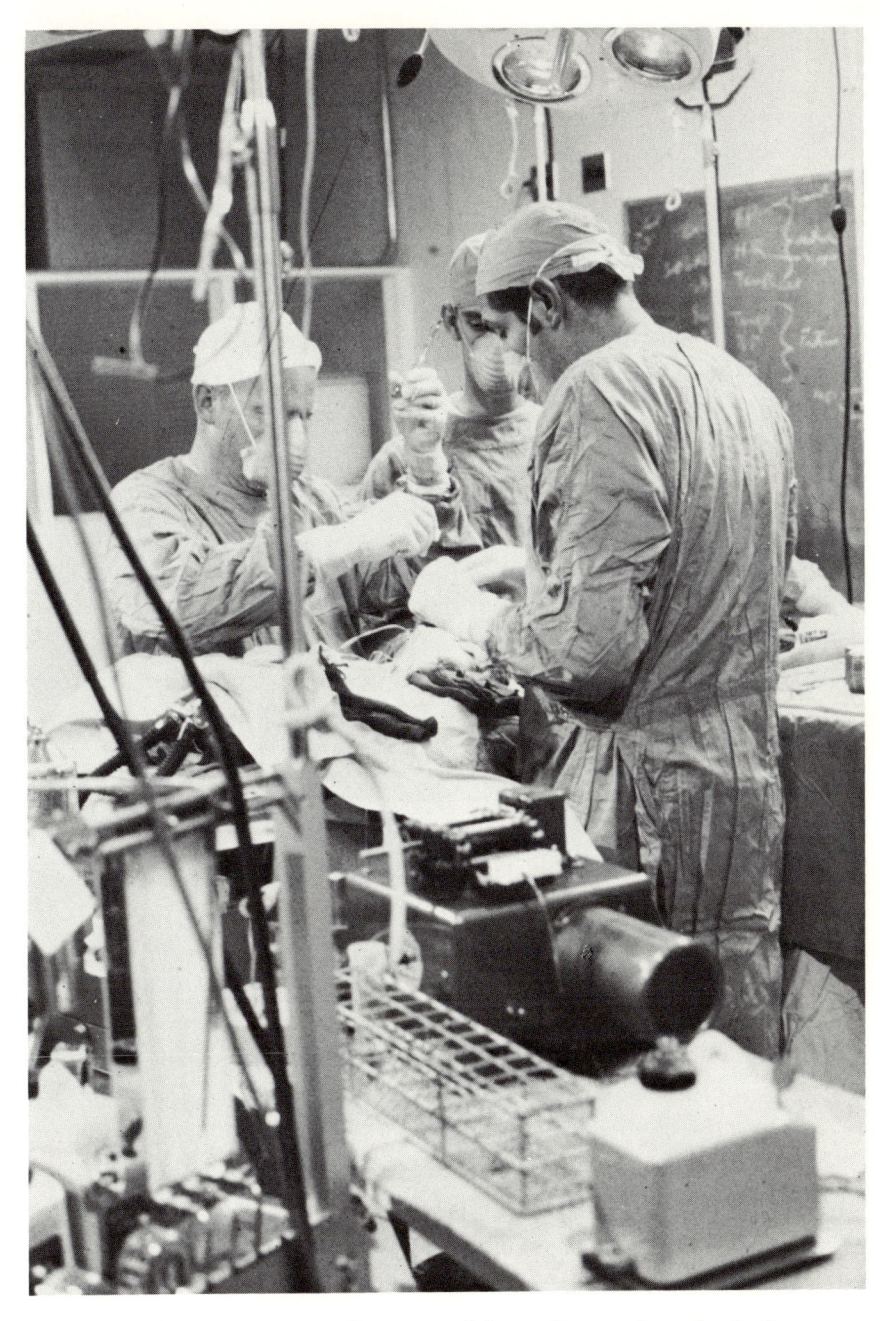

A monkey fetus is temporarily removed from the womb and wired with a variety of sensors. The effects of drugs given to the mother, can then be closely observed in the fetus. (*Don Blauvelt*)

various laboratories. But always the embryonic subject has been an experimental animal. This is not the case with the larger and thus more easily manipulated fetus. Here both monkey and man have been the subject of fetal surgery. At the Oregon Regional Primate Center a team headed by Dr. Richard Behrman opens the pregnant monkey's womb, partially removes the fetus, and inserts a number of catheters, electrodes, and other monitoring devices to various points on the legs and head. The object is to be able to get a close-up look at brain function, heart function, blood flow, and a number of other vital activities that occur in the fetus during pregnancy. At present we now know little or nothing about these functions. The effects of drugs and of viruses and other disease agents that might cross the placental barrier can also be studied in the womb, for these fetuses survive the surgery quite readily and go on to full-term live births.

With humans, the experimental aspect of fetal surgery is ignored, and it becomes only a desperate last-gasp effort to save a dying fetus. At Columbia's Presbyterian Hospital, Dr. Karlis Adamsons and Dr. Vincent J. Freda have intervened surgically to try to save fetuses that are too ill with anemia to survive even the standard intra-uterine transfusion. The surgery involves partially delivering the fetus, by Caesarean section, and then placing a catheter in the fetal abdominal cavity. The fetus is then returned and the uterus closed to allow intra-uterine growth to continue. The catheter becomes the fetal lifeline, through which life-saving blood is piped as needed. This incredible procedure has been undertaken twenty times and, equally incredible, it was accomplished without triggering premature labor. Such desperation attempts have managed to keep some of the fetuses alive for a few weeks longer than they would have survived without intervention, but only a few were actually brought to term. Dr. Freda was able to keep one doomed fetus alive in the womb for two months by means of an implanted catheter, but this is the longest time that the technique has been successful. The problem has been that most fetuses were already too sick to be saved when help reached them. Dr. Adamsons has also tried to cut out fetal cancers and kidney obstructions and to repair fatal diaphragmatic hernias—holes in the muscular sheath that separates the chest cavity from the abdomen.

Fetal surgery, even in the 21st century, is not likely to become a routine event. Most fetal problems can be held in check by the placenta until after birth, when surgery can then be more easily accomplished. But the use of fetal surgery for those cases that cannot wait is becoming more and more a reality. It is also one of the great challenges facing science as we approach the 21st century. "There are only three areas in the world left to explore," declares Dr. Sidney Gellis of Tufts University, "outer space, the sea, and the womb. We are really at the last frontier."

4 | Standing Room Only

There exist in the world today some three and a half billion people. Every time your pulse throbs, still another person is added to that total. In another thirty-eight years, that kind of arithmetic will have doubled the population of the earth. Even if birth control were to become the universal rule from this moment on, it would be next to impossible to alter the fact that the 21st century will dawn over a population of seven billion people.

The numbers demographers must use to describe future populations have begun to rival those astronomers use in measuring interstellar distances. And it is these astronomical numbers that may end man's glorious adventures. It appears that the most fearful detonator may be biologic rather than atomic, and the explosion that renders the planet uninhabitable will in all likelihood be a population explosion.

Man has spent approximately one million years on this earth, and for most of that time his numbers remained fairly constant. Infant mortality rates were extremely high, food was always in short supply, and disease killed off young and old in great plagues and lesser epidemics. Not too many people survived to the child-bearing age. Under such conditions it took about thirty thousand years for the population of the earth to double. At the time of Christ the world contained perhaps three hundred million people. It took another seventeen hundred years to double that figure. Then the rate picked up as the population doubled again in two hundred years, and yet again in less than one hundred years. At this rate of growth, the population will double again every thirty to forty years. What worries the experts, however, is that the rate is not constant, it is increasing.

"This marked rise in population," notes Dr. Roger Revelle, director of the Harvard University Center for Population Studies, "really started about 1940. At that time the Earth's population was doubling about once every seventy years, about a one per cent increase a year; now it's doubling about once every thirty-eight years, about a two and a half per cent increase per year. And it's this rate of change that's really the frightening and disturbing aspect of human populations."

"Had this rate existed from the time of Christ until now," declares Dr. Clement L. Markert, chairman of the department of biology, Yale University, "the world population would have increased in this period by a factor of about 7×10^{16}. In other words, there would be about

In poor nations, such as Korea, overpopulation wipes out all economic gains. To cope with the rising population, teams now cover the Korean countryside, dispensing birth-control information and devices.

Every communications device is used to get the birth-control message across. Here, a singing commercial lauds the virtues of family planning in Korea.

twenty million individuals in place of each person now alive, or one hundred persons to each square foot.

"Calculations of this sort demonstrate without question, not only that the current continued increase in the rate of population growth must cease, but also that the rate must decline again. There can be no doubt concerning this long-term prognosis; either the birth rate of the world must come down or the death rate must go back up. There is no escape from balancing the equation of births and deaths. Nature's processes are certain, ruthless, unemotional, impartial, and as sure as death. If man does not control his numbers voluntarily, then we can be sure that the traditional methods of famine, pestilence, and war will provide us with highly unpleasant limiting controls."

That nature does possess an awesome arsenal of weapons for population control has been demonstrated in every species. In man, the great leveling influences have been disease and food supplies. There were times in man's history when infectious diseases, such as bubonic plague, killed one of every three people on the face of the earth. Famine has long been the great lid holding down Asia's teeming populations. Natural disasters that destroyed crops or rains that did not come to allow crops to be raised have in the past had deadly consequences for India, China, and other populous but poor nations. Millions would die of starvation, falling in the streets to be carted away in wagons and burned on some communal pyre or interred in a mass grave. But both of these inexorable population levelers have suddenly been swept away. Wholesale death from epidemic disease has almost disappeared as a contemporary event. A combination of immunology and sanitation has greatly limited the spread of infectious disease. A similar combination of chemistry and environmental control has increased the food supply enormously to the point where famine, even in most of the desperately poor nations of the world, is now reduced to the nature of a threat and not an ever-present happening.

The result is, as has been noted, an almost astronomical increase in the human population that has thrown off for the moment at least all natural brakes. In other species, nature has demonstrated a flexibility of approach to compensate for factors that no longer exercise a governing effect.

In the absence of disease and nature's great balancing mechanism, predators, only the supply of food would seem to be effective in limiting animal populations. Most often those three factors act in concert to maintain an equable population of all species sharing an area. But occasionally an abnormal situation arises, and one or more of the governing factors is removed. This was the case on James Island, in Chesapeake Bay off the Maryland coast. In 1916 someone stocked the 280-acre woodland island with four or five deer. With no predators and a more than adequate food supply, the population on James Island soared until it reached a density of about one deer per acre.

To a deer herd, this represents extreme crowding. The effects of such crowding were startling. Within the first three months of 1958 more than half the deer died suddenly and inexplicably, for nothing else had changed; the food supply was still adequate for the entire herd and there were still no predators on the island. The next year the wholesale deaths ceased, and the population stabilized at about eighty deer.

The death of the deer occurred before the eyes of a scientist named Dr. John J. Christian, an endocrinologist now at the Albert Einstein Medical Center in Philadelphia. Christian performed autopsies on the deer before and after the population had stabilized. His findings ruled out malnutrition, epidemic disease, and cold. Most of the dead deer had grossly enlarged adrenal glands and pathological symptoms of chronic kidney disease. The trigger, according to Dr. Christian, was social pressure.

In subsequent experiments on rats, woodchucks, mice, deer, and other animals in a variety of laboratory and natural settings, Dr. Christian has found essentially the same phenomenon at work. Increasing populations produced competition and other social pressures that caused changes in the endocrine systems of individual animals, changes severe enough to kill them, even when sufficient food and water and ostensibly sufficient space were available.

"What we have seen," explains Dr. Christian, "seems to be a feedback mechanism designed to assure survival. It appears to be a safety check to stop populations from exhausting the resources of their environments and it seems to start working long before food becomes a crucial factor.

"Although our experiments have been with animals, we believe the same mechanism, with certain variations, of course, might be at work in human populations."

Aberrant social behavior especially seems to be the controlling mechanism for social beings. Among the highly organized social colonies are those of rats. The rat is perhaps the most successful creature on earth; his only real competitor in terms of cleverness, destructiveness, and adaptability is man himself. Studies of the rat, have revealed uncomfortable parallels to man. Consider the effect of population pressures on the social behavior of rats. At the National Institute of Health in Bethesda, Maryland, Dr. John B. Calhoun penned a group of wild Norway rats in a rat paradise—a quarter-acre enclosure with ample food and water, material for nests, and no natural predators or disease allowed.

When mice become overcrowded a number of bizarre events take place. Mothers either abandon their young or overprotect them. Males fight each other and eat the young. The ultimate result is a soaring increase in infant mortality until the population decreases to a number that the environment can sustain.

The experiment had been so constructed that there could be only one governing control on the ultimate population of the colony—social behavior. After twenty-seven months the population within the enclosure, had stabilized at one hundred and fifty adults. "Yet," noted Dr. Calhoun, "adult mortality was so low that five thousand adults might have been expected from the observed reproductive rate. The reason this larger population did not materialize was that infant mortality was extremely high. Even with only one hundred and fifty adults in the enclosure, stress from social interaction led to such disruption of maternal behavior that few young survived."

With an obviously erratic behavior pattern established by increasing

Psychological pressures are also noted in overpopulated rat and mice colonies. Here the mice huddle in one corner of the pen and devote most of their time to just eating and sleeping. Ninety-five per cent of the adults in this crowded colony have become sterile.

population pressures, Dr. Calhoun resolved to determine the specifics of the societal breakdown within a rat colony under controlled conditions. Using a domesticated strain of the Norway rat in the laboratory, he allowed the population to increase to about twice the size that could normally fill the available space. The effects of such crowding were then observed for sixteen months on six different rat colonies.

A composite picture of life within the six crowded colonies was drawn. "The consequences of the behavioral pathology we observed were most apparent among the females. Many were unable to carry pregnancy to full term or to survive delivery of their litters if they did. An even greater number, after successfully giving birth, fell short in their maternal functions. Among the males, the behavior disturbances ranged from sexual deviation to cannibalism and from frenetic overactivity to a pathological withdrawal from which individuals would emerge to eat, drink, and move about only when other members of the community were asleep."

The social organization of the colony suffered as much as did its individual members. One of the most disastrous effects was the development of what psychologists call a behavioral sink. The colony was maintained in a series of four interconnecting pens, but as many as sixty of the eighty rats in the population would jam into one pen at feeding times. Soon the rats were spending all of their time in the feeding pen, leaving the other pens occupied only by those few rats that could not force their way into the feeding pen.

"Eating and other biological activities were thereby transformed into social activities in which the principal satisfaction was interaction with other rats," explains Dr. Calhoun. "In the case of eating, this transformation of behavior did not keep the animals from securing adequate nutrition. But the same pathological 'togetherness' tended to disrupt the ordered sequences of activity involved in other vital modes of behavior, such as the courting of sex partners, the building of nests, and the nursing and care of the young. In the experiments in which the behavioral sink developed, infant mortality ran as high as ninety-six per cent among the most disoriented groups in the population."

That man might have his own behavioral sink into which he is rapidly slipping is a question to contemplate. It seems obvious that the breakdown in social order, the violence and mental stress that have become a very real part of our everyday lives, is indicative of some pressure that we cannot cope with. Perhaps there is even an analogy with the rat in a shocking upsurge in child beating and brutal maltreatment that has no parallel in modern history. Increasingly, doctors are treating children for "falls," and "burns" and other "accidental injuries" that are accidental only in the parents' description of them.

So prevalent has the phenomenon become, doctors have given it a special label, "the battered-child syndrome." Its victims are usually under three years of age. One in ten dies, while fifteen per cent of the survivors suffer permanent brain damage.

Is there a parallel to the actions of Dr. Calhoun's overcrowded rats?

"The middle-pen females," he reported, "lost the ability to transport their litters from one place to another. They would move only part of their litters and would scatter them by depositing the infants in different places or simply dropping them on the floor of the pen. The infants thus abandoned throughout the pen were seldom nursed. They would die where they were dropped and were thereupon generally eaten by the adults."

The experiments of Calhoun, Christian, and dozens of other biologists in laboratories throughout the world have always demonstrated that there is a natural governor, expressed in terms of social behavior, that works to limit the population of any single species. It is also apparent that all biological systems studied function according to the same general laws. This would indicate that man is no exception and the same social pressures might be expected to curb his growing numbers. Despite this evidence, many students of the population growth view the problem somewhat simplistically as a race between reproduction and production. The development of new sources of food, of increased crop yields and exotic new foods, fish farming, and even the creation of vast new supplies of synthetic foods are seen as a number-one priority in the next few years.

There is no question that a sharply increased food supply is needed. Two thirds of the world's population go to bed hungry every night. Simply to keep pace with the burgeoning population, the world production of proteins will have to be increased fivefold by the year 2000. Most of the increase must be accomplished in what have been euphemistically labeled the underdeveloped nations of the world. These are, in plain words, the poor countries of Latin America, Africa, and Asia, where two thirds of the exploding population will be born between now and the 21st century. The likelihood of those countries' accomplishing such a monumental increase in their food supplies is small indeed.

As a result, many experts are predicting famine throughout the underdeveloped "Third World," as the French call it, in the next fifteen to twenty years. This is the "crossover" point, when all surplus food produced in the rich nations of the world will still not be enough to meet the demands of the chronically food-short poor nations.

The significance of these food handouts can be appreciated by a hypothetical situation *Fortune* magazine posed in an article on hunger.

Geochemist Harrison Brown of the California Institute of Technology calculates with horror the idea that if our technology were applied efficiently to the problem of overpopulation we could provide for 200 billion people on the earth.

Using a Department of Agriculture standard of 2500 calories a day as adequate, the magazine raised the following point: "If India had distributed its supply of food as far as it would go even at only a 2300 calorie level, forty-eight million out of that country's four hundred and eighty million in that year (1963) still would have been left totally without food."

Even without observing such niceties as a barely adequate diet, the prospects for wholesale famine in the Third World are depressingly good.

Faced with such a dilemma, with millions of people starving ("We shall see them doing so upon our television sets," C. P. Snow told a shocked audience at Westminster College in Fulton, Missouri, recently), the rich nations will undoubtedly bestir themselves. Immense infusions of technology in the form of fertilizer, fish-protein concentrate, and laboratory-generated proteins will slow somewhat the pace of starvation.

With the proper application of technology and singlemindedness of purpose, there seems almost no limit to the number of people the earth might support. "If we were willing to be crowded together closely enough, to eat foods which would bear little resemblance to the foods we eat today, and to be deprived of simple but satisfying luxuries such as fireplaces, gardens, and lawns, a world population of fifty billion persons would be not out of the question. And if we really put our minds to the problem we could construct floating islands where people might live and where algae farms could function, and perhaps one hundred billion persons could be provided for. If we set strict limits to physical activities so that caloric requirements could be kept at very low levels, perhaps we could provide for two hundred billion persons," says Harrison Brown, geochemist at the California Institute of Technology.

Such ghastly predictions are meant more to dramatize the problem than as a reasonable prospect. Still there is no denying the mighty power of technology properly applied. It could in fact lead to what has been described as the "utterly dismal theorem."

"This is the proposition," says economist Kenneth E. Boulding, "that if the only check on the growth of population is starvation and misery, then any technological improvement will have the ultimate effect of increasing the sum of human misery, as it permits a larger population to live in precisely the same state of misery and starvation as before the change...."

Boulding uses Ireland as a case in point. The population of that country in the late 17th century was approximately two million people, almost all of them living in abject poverty and misery. Then came the introduction of the potato, which raised the food yield per acre to dizzying heights. The population also soared, from two million to eight million by 1845. "The result of the technological improvement, therefore," points out Boulding, "was to quadruple the amount of human misery on the unfortunate island."

Boulding, of course, refers to the disaster of 1845 when the potato crop failed. "Two million Irish died of starvation; another two million emigrated; and the remaining four million learned a sharp lesson which has still not been forgotten. The population of Ireland has been roughly stationary since that date, in spite of the fact that Ireland is a predominantly Roman Catholic country. The stability has been achieved by an extraordinary increase in the age at marriage."

The Irish may have learned, but that is not to imply that the rest of the world will be equally pragmatic. Cultural patterns and religious dogma ingrained by centuries of use stand between realistic population control and disaster.

In the poor, primarily agricultural nations of the world, large families have been the rule for time out of mind, a hedge against the inordinately high death rate. In India today one of every five children dies before the age of five. This is a great improvement over the death rate of previous generations, when seven of every ten children would not live to adulthood.

The improving statistics, however, are meaningless to an Indian villager, who is himself one of two surviving children in an original household of eight. To him, and those of his generation, children are an absolute necessity.

"They want to have the chances very much in their favor that at least one of their sons is going to grow up to be a man," explains Roger Revelle. "In the first place, they're liable to starve to death in their old age unless it is so. In the second place, the meaning of life, the very justification for all the misery and poverty that people live in is somehow to perpetuate life, to carry on, and to leave something behind. This, in the poor countries usually means leaving a son. The Hindus say, 'I need a son to light my funeral pyre.' This means a great many things more than it says."

Harrison Brown points out that "asking an Indian farmer to practice birth control is like asking an American worker to tear up his social security card."

A similar problem exists for the world's half billion Catholics, whose sexual habits are, in theory at least, dictated by a celibate Pope. The Pope, acting after what was probably an agony of deliberation, re-established in his encyclical "On Human Life," that no form of birth control other than the rhythm method may be practiced by Catholics. Asking the author of such an encyclical to accept other forms of contraception as a matter of individual conscience for Catholics, and for that matter anyone else, requires in effect a theological dichotomy similar to the cultural dislocation that is called for from the Indian farmer.

And yet, this is the nub of the situation, the one point on which everything must turn or nothing will. Technology with its pills, loops, and other contraceptive devices stands ready, but custom and Church are not. What is needed is a massive shift in traditional ideas.

"If we can establish sufficiently that no married pair ever have a child that they do not want, whether this unwanted child is a first, third, or tenth, we would go a long way towards setting up a climate of opinion in which the idea of choice of whether or not to be a parent ever, once, or several times moves into the center of public consciousness," says anthropologist Margaret Mead of Columbia University.

Effective means of contraception have always been available, even before the recent revolution in biotechnology. Even the most primitive of agricultural societies found it necessary at one point or another to limit their fertility. One means was by prolonged nursing. Nature, recognizing the extreme demands of human young, has somehow arranged it so that a nursing mother stands little chance of becoming pregnant.

One of the more popular of primitive population-control techniques was infanticide, in the guise of either religion or plain displeasure with the baby. Abortion is another of the long-familiar and still practiced methods of birth control.

In some societies neglect and abuse of female children killed off many prospective mothers, while social conventions and sexual taboos limiting the frequency of intercourse all contributed to a fairly effective form of population control.

In the more advanced societies, in Europe, for example, before the advent of contraceptive devices, nations limited their populations simply because their people wanted to do so.

"They depended on just a strong motivation not to have too many children," says Roger Revelle. "They postponed marriage, they practiced the rhythm method, they practiced withdrawal, they did whatever was necessary not to have too many children."

The tremendous drop in mortality rates all over the world simply means that in those nations where the birth rate is highest and food and technology are in the shortest supply, the traditional methods of birth control, geared for the most part to being put in effect only after several children have already been born, have become totally inadequate.

Populations continue to increase despite the fact that nature herself seems determined to make conception an extraordinarily difficult event, one that can take place only within a period of from three to six days every month. The chain of events that leads to that time begins with the female menstrual period, when the uterus sloughs off its endometrial lining laced with blood vessels. This flow of the menses is caused by the changed hormone balance that also triggers a biochemical signal to the pituitary, the so-called master gland, at the base of the brain. The pituitary releases a hormone called FSH (follicle stimulating hormone) into the bloodstream. Once the FSH reaches the ovaries, it does exactly what its name says—it stimulates the growth of a sac called the Graafian follicle, within the ovary.

Every normal female is born with about fifty thousand of these egg-carrying follicles in her two ovaries. Each month after puberty the follicle swells as the egg ripens inside the ovary. Finally, it erupts, spilling the egg onto the surface of the ovary. This is ovulation, the beginning of the critical period of several hours during which the ripened egg is susceptible to fertilization. First, the egg must be swept up by one of two prehensile funnels called the Fallopian tubes. Then it must travel down the tube and into the uterus. It is a distance of only four or so inches, but the journey takes three to six days.

If intercourse takes place and if one of the millions of male sperm that are deposited in the vagina finds its way through the cervix, the narrow hallway leading to the uterus, and into the Fallopian tube, and penetrates the barrier around the egg cell to its interior, then and only then will fertilization occur.

While the frenetic activities of sperm and egg go on, the marvelously intricate endocrine system orchestrates the flow of hormones. The

Graafian follicles, having surfaced the egg from the ovary, now spill estrogen into the blood stream.

The rise in the estrogen level is noted by the pituitary gland, and it eliminates the production of FSH, thereby preventing the development of any more eggs, and accelerates the production of a hormone called LH, leutenizing hormone. The LH acts as a chemical guardian from the moment of ovulation until the fertilized egg is implanted in the lining

At the Weizmann Institute in Israel Dr. M. C. Sheleznyak holds a rat while a colleague injects a fungus extract designed to prevent the implantation of a fertilized egg in the womb. The compound is injected into the mice for up to five days after fertilization has occurred. In ninety-nine per cent of the cases the fertilized eggs simply disintegrate. This may be the long-sought postcoitus contraceptive—the "morning-after pill."

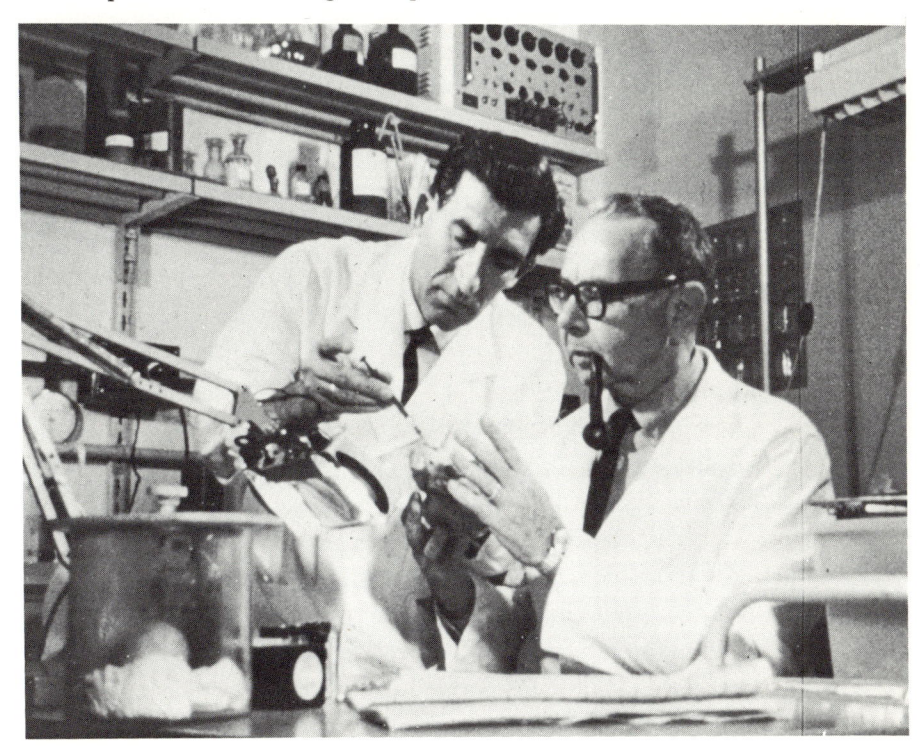

of the womb. It encourages the growth of a spongy mass, the corpus luteum, within the ovary. The corpus luteum, in turn, manufactures yet another hormone called progesterone, which triggers changes in the breasts, preparing them for milk production, and in the endometrial lining of the uterus, which grows thicker and softer, and becomes rich in nourishing blood vessels in anticipation of the implantation of the fertilized egg.

If in fact the egg has been fertilized and is implanted, the intricate biochemical switching of hormones on and off now ceases and a supply of progesterone is provided by the placenta itself while the complex process of embryonic growth takes place. In the event that the egg has not been fertilized the flow of progesterone ceases as the corpus luteum comes to the end of its two-week life span. This is followed by the sloughing off of the deteriorated, unfertilized egg along with the endometrial lining. The entire cycle then begins anew.

So complex is the process, it would seem that birth control would be a relatively easy affair. And so it is; any interruption, alteration, or suppression of any single link in the intricate chain will prevent conception. But again, the over-all rules that govern biological systems prevail. The reproductive system has a great balance to it. Enormous numbers of sperm are available, even though only one egg need be fertilized, and the most fertile of women rarely are capable of having more than ten or twelve children during a lifetime. Checks and balances are the rule, one system pulling against another, complexity coupled with endless repetition, so that the ultimate goal of biology, survival of the species, may be achieved.

The most effective means of preventing overpopulation, and at the same time permitting intercourse to take place unimpeded, is the birth-control pill. But no one knows precisely how it works. "The pill," as it has become known, is composed of estrogen and progestogen, a synthetic version of progesterone. Taken after menstruation for a period of twenty days, the pills furnish artificial hormones that seemingly fool the body by mimicking the levels that normally shut down the pituitary. Noting this, the pituitary fails to release the FSH needed to develop an egg within the ovary. The artificial hormones also stimulate the lining of the uterus as if it were preparing to receive an egg. After the twentieth pill has been taken the hormonal level drops back to what would be the normal level for a nonpregnant woman just before menstruating. The endometrium is sloughed, giving a menstrual flow, and the artificial cycle can then be repeated.

In addition to its alteration of the hormonal level, the pill seems to effect other changes in the complex reproductive cycle. The endometrial lining does not thicken as much as normal, as if it would not accept a fertilized egg even if one were to find its way there. Within the cervix, too, changes occur that seem to prevent conception. The viscosity of the ever-present mucus that lubricates the cervix changes to form an almost impenetrable barrier to the sperm deposited in it.

The changes are such that the pill appears to be, after ten years of use, one of the most effective drugs ever devised. And the cost has been dropping, so that eventually it may become practical as the birth-control method for even the poorest of nations. Some eleven million women outside the United States are already taking the pill, and each day brings new users. But the pill does have its drawbacks. In some women the pill causes weight gain, and fluid retention, which produces a slight swelling of the breasts, finger joints, and limbs. A British study found a slightly increased risk of thromboembolic disease for women taking birth-control pills. Although the possibility that the pill might cause cancer has been discussed since birth-control pills first came into general use, researchers have not yet been able either to confirm or to deny a relationship between the pill and cancer of the breast or uterus, the two sites thought to be affected by hormonal contraceptives. There remains, however, the possibility of a latent effect, of a long period of time between the administration of a carcinogen and the appearance of a recognizable cancer. Not enough time has elapsed to rule out this possibility.

For some women, a serious shortcoming of the pill is the rather exacting schedule that must be followed to make it effective. Even illiterates can learn to count to twenty, but every woman, regardless of education, is prone to a human lapse. When the penalty of such a moment of forgetfulness is pregnancy, the price seems exorbitant.

For those who feel that the price of both pill and pregnancy is too high, a mechanical method of birth control, the intra-uterine device, or IUD, may be more acceptable. According to legend, the IUD was developed originally along the ancient caravan routes of the desert.

At the beginning of a long trek, the Arab drivers would place a round, smooth stone in the vagina of a female camel. In some way, this kept the animal from becoming pregnant and thus for a time unusable as a beast of burden in the caravan.

In the 1920s a German physician adapted the idea for the human womb by replacing the stone with a ring of silver cushioned with surgical silk. Unfortunately, many gynecologists feared that the procedure would give rise to severe infections, so the idea was abandoned, only to be picked up again in the 1960s, with the introduction of plastic coils, loops, and curlicues that could remember their original shapes after being temporarily straightened out for easy insertion into the womb. Doctors are uncertain about how the IUDs work, and there is no question that they are less effective than the pills. They do possess two attributes the pill does not. One is their cost, which is pennies, compared to the approximately two dollars a month for the pills. The second is that, once inserted, the IUD can be forgotten about in most cases.

The great drawback is the fact that IUDs must be inserted by either a physician or a highly trained nurse. In inept hands, they are quite dangerous, capable of puncturing the wall of the uterus, and, of course, not at all effective if inserted improperly. Not all women can retain the IUDs; about seven per cent expel or displace the device as a result of normal uterine contractions. More often than not the woman notices the expelled device and knows that it must be reinserted. Otherwise, the first sign that expulsion has occurred is usually pregnancy. This gives the IUD a ninety-seven per cent effectiveness the first year it is in use, which increases to ninety-eight per cent thereafter.

"With its higher failure rate," says Dr. Alan Guttmacher, president of Planned Parenthood–World Population, "the IUD may not be satisfactory for your wife, but it may still be good enough for a public-health program in a developing country."

But is it really? Against a biological time bomb that is ticking away at a frightening rate, the limitations of the IUD may seriously restrict its usefulness. What is needed is an absolutely effective, extremely simple, and inexpensive contraceptive that will fit within the cultural and aesthetic framework of the sexual act.

One approach is now under development by Dr. Sheldon Segal,

director of the Biomedical Division of the Population Council at Rockefeller University. Dr. Segal thinks that the real need is for a long-acting method, something that can be taken once and then forgotten for a year or twenty years. His answer is a silicone-rubber capsule filled with progesterone. The capsule is implanted under the skin and releases a precise amount each day into the bloodstream. Because the hormone enters the bloodstream at a constant rate, the hormonal highs and lows measured in blood levels when a daily pill is taken are prevented. This means that far less hormone, as little as one tenth the amount used in

Among the many birth-control methods now being explored is a long-term version of the pill. This silicone-rubber capsule is filled with birth-control hormones that are released automatically into the bloodstream. The capsule is fitted into a needle for injection beneath the skin.

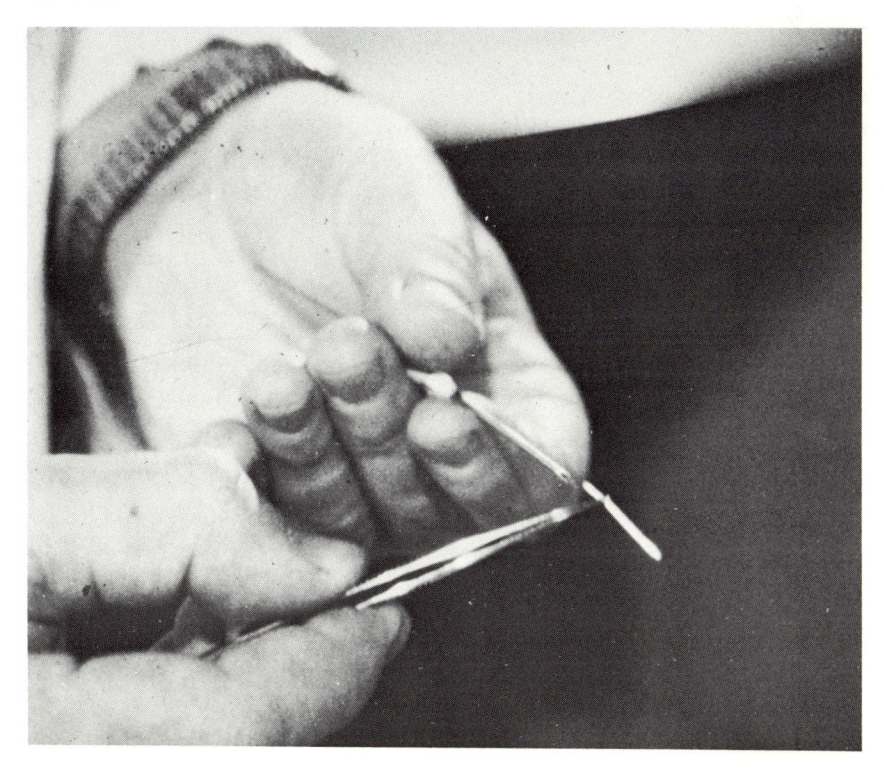

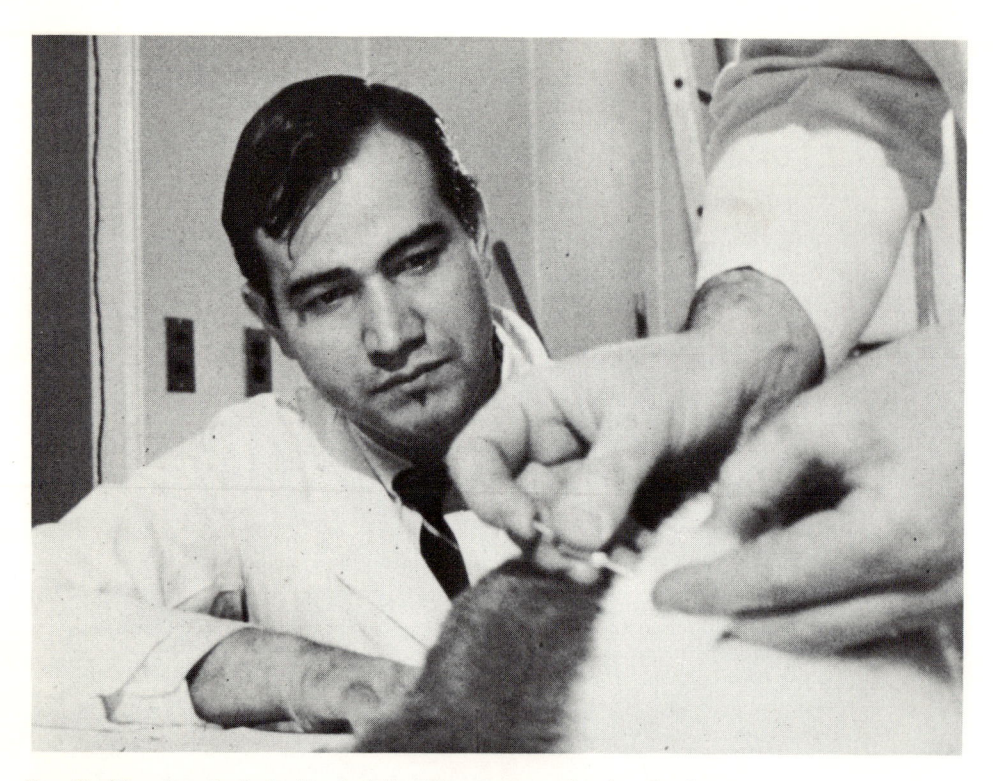

Dr. Sheldon Segal of the Rockefeller University, who heads the team working on a long-acting birth-control capsule, inserts an experimental capsule into a rabbit, while a technician looks on.

present-day birth-control pills, is needed. This will cut both costs, and possible side effects.

Dr. Segal has measured the amount of hormone released through the pores of the silicone capsule each day and thinks that a single capsule implanted under the skin could theoretically be effective for twenty years.

Dr. Segal now thinks that the capsule could fit inside a hypodermic needle, so that it could be injected into the patient, rather than implanted surgically.

The long-term implant, or "time capsule," could be a reality in just a few years. To reverse its effect, a woman would simply have to request that it be removed.

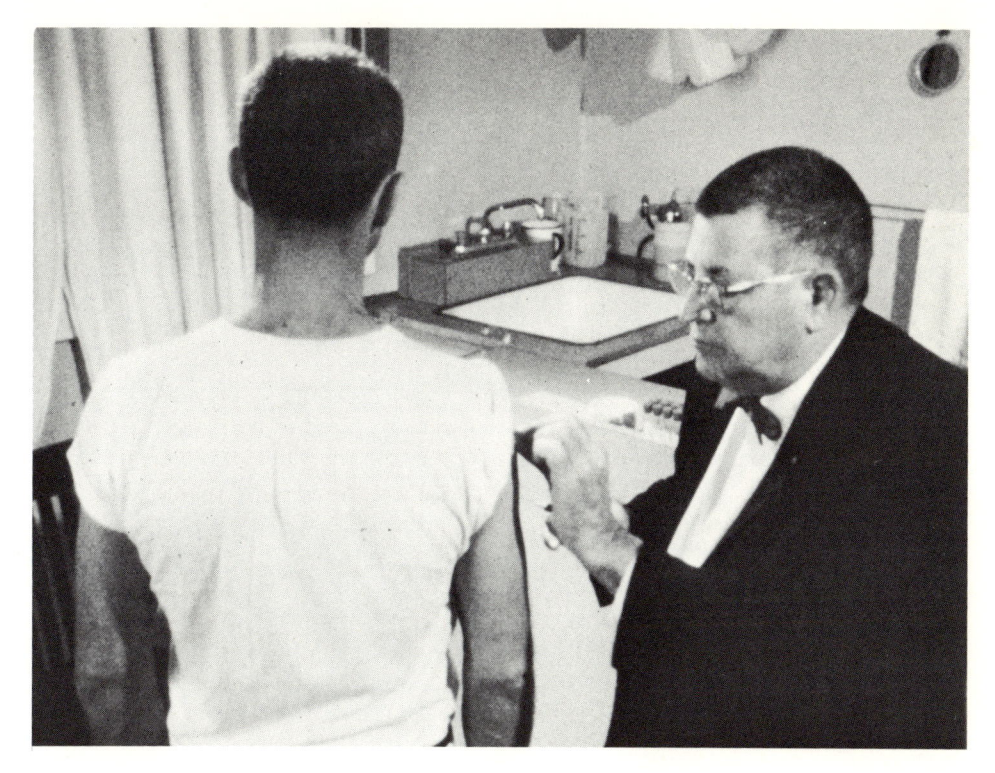

One approach to the problem of overpopulation is to develop a birth-control pill for men. Hormones that inhibit sperm production are tested on volunteers at Alabama's Kilby State Prison.

"Right now," says Segal, "if a woman wants to prevent pregnancy, she has to take the positive step of using a contraceptive method or to abstain from intercourse. This requires positive decision and action. Women of the 21st century who may have such capsules implanted could make the decision once to have the capsule implanted, and then it would take a positive step and a positive decision to have a child. She would have to make the decision to go and have the capsule removed."

Other researchers, convinced that women, rich or poor, socialites or village illiterates, simply cannot count at all, are developing a pill that is taken every single day. That way, they insist, it becomes a rote maneuver, as habitual as washing, and there can be no mistakes. Still other researchers are seeking a morning-after pill, one that will not interfere with ovulation but will simply prevent the egg from being fertilized or implanted in the lining of the womb. In theory such a pill would be effective if taken any time within three to six days after intercourse (the normal time it takes the egg to travel from ovary to uterus).

There is no shortage of methods to prevent conception. What is lacking is a willingness to cooperate on the part of those who will have children and, even more important, a sense of the urgency, of the terrible danger of the situation, on the part of government and religious leaders everywhere.

"There are already too many people in the world," C. P. Snow declared. "Within a generation, there will be far too many. Within two or three generations—unless we show more sense, good will, and foresight than men have ever shown—ordinary human hopes will have disappeared."

And with them all hopes for the 21st century. If the dire predictions and baleful warnings of the scientists and intellectuals of the world are not now heeded, the quality of life in the 21st century will be marked as among the most wretched in human history. But if the population bomb can be defused, life in the 21st century might well be worth living for all.

5 | Man-Made Man

While one group of scientists struggles with the problems of birth control and a population explosion that threatens to overwhelm the world, still another group is developing remarkable techniques to control death. The paradox is further heightened by the role of nature, which has carefully evolved complex processes to protect the species by championing birth and death at the same time. Thus, the thrust of biological forces is in the favor of the species—mankind—and heavily weighted against the individual—man.

Disease, illness, and injury have always taken their toll of man, and the complex structure of the body itself seems to have but a limited span of existence built into it. For all its elegant complexity, the human body is a delicate vessel in which to carry life. The frailty lies in the organs and systems that power and protect the body. Man cannot live without brain, heart, liver, and one each, at least, of his two kidneys and two lungs.

Until the mid-20th century, doctors could hope only to repair or remove diseased or damaged organs. But now we stand on the verge of a great change. We are learning not only to repair but also to replace the vital organs, the once irreplaceable mechanisms of life.

Each of the vital organs now seems either repairable or replaceable. Already kidney transplants have become so routine that they are no

Dr. Adrian Kantrowitz, chief of surgery at Maimonides Medical Center in Brooklyn, N. Y., and one of the world's leading transplant surgeons.

longer even announced to the press. Even the transplantation of a human heart is no longer considered front-page news, and the problems of transplantation now focus upon rejection and supply of organs, upon developing mechanical substitutes for living tissue, and upon sorting out the complex legal and moral problems that have been created by the surgical miracles that are fast becoming a commonplace.

"I wouldn't be surprised," says Dr. Adrian Kantrowitz, chief of surgery at Brooklyn's Maimonides Medical Center, "if I came back in the year 2000, if I walked around a hospital and they showed me patients with transplanted kidneys, with transplanted lungs, with transplanted hearts, with transplanted pancreases, stomachs, intestine—all these things have been done in the experimental laboratory."

What has been done most often in clinical practice, however, has been the transplantation of the kidney, and that organ will probably be the one most commonly transplanted even in the 21st century. The primary reason is its availability. Every normal human being is born with two bean-shaped, fist-sized kidneys. Their importance to the body is so great they they are tucked away in the lower back and padded in heavy layers of fat for additional protection. And they come in pairs, one kidney providing a back-up or emergency system for the other. The healthy kidney comes close to being the most efficient of all organs. Packed within its five ounces are a million invisible filtering units, called nephrons, whose coiled and twisted tubules and tufts of blood vessels, called glomeruli, would stretch to a combined length of more than 140 miles. Through this network of infinitesimal canals flow, on the average, eighteen hundred quarts of blood every day. The two kidneys process almost three times the body's weight in water and salts every twenty-four hours. About one and a half quarts are excreted as urine; the rest is returned to the main circulation.

Should the exact chemical formula of the fluids and salts in the body alter from the norm, the glomerular tufts and tubules that make up the nephron immediately readjust the vital balance. Too much of one chemical, and the nephron funnels it out in the urine. If there is too little, the tubules collect and store it until the proper proportions are restored.

This liquid formula constantly maintained by the kidneys is virtually identical to the chemical make-up of the seas that covered the earth when life began. Then, the organisms that inhabited the earth needed no kidneys, for their internal fluids exactly matched the seas in which they dwelled. But as the waters receded—over millions of years—some sea creatures adapted to the land. To survive outside their natural habitat, they developed, among other organs, a system to maintain a constant chemical balance within the body that was quite similar to that of the sea. One of the most important of the organs that enabled those early pioneer sea creatures to evolve into land-dwelling mammals, and eventually into man, was the kidney.

The final product of all the filtration, the total of millions of tiny drops emerging from each of the nephrons, streams from the kidney through an inch-wide funnel into a narrow, foot-long tube called the ureter and into an extremely elastic pouch called the urinary bladder. When necessary, this marvelously distensible bag can expand to hold a quart of fluid. But normally, as little as a cupful of urine puts enough pressure on the bladder walls to trigger a nerve signal to the brain. The brain then tells us it is time to urinate.

So efficient is the urinary system that when disease or injury does not interfere, the body could get along on only one fifth of the system's capability. But disease especially does interfere, so much so that a hundred thousand people die of chronic kidney disease every year in the United States alone. But that same marvelous efficiency of the healthy urinary system makes it possible for a healthy person to donate a kidney to another and never miss it.

Research with an eye toward transplanting kidneys began a generation ago, in 1946, when Dr. Jean Hamburger, of the Hospital Necker in Paris, began experimenting with dogs, and then transplanted the kidneys of guillotined convicts into dying human patients. The transplants never worked, for within a few days the bodies of the gravely ill patients always rejected the organ that would save their lives.

Then during the Christmas of 1952, a sixteen-year-old plasterer's apprentice named Marius fell from a scaffolding to a Paris street. The boy was rushed to a hospital where examination revealed a crushed and hemorrhaging kidney. To prevent him from bleeding to death internally, the damaged kidney was perfunctorily removed. This procedure was standard, since man functions quite well with just one kidney. To the surgeons it had been an almost routine case, with no complications.

Five days later, the doctors' certitude was badly shaken. Marius had not urinated since entering the hospital. A hasty physical examination revealed that the kidney that had been removed was his only kidney.

Death was inevitable. In a futile hope, Marius was sent to the Hospital Necker and Dr. Hamburger for a never-before-successful kidney transplant.

The barrier was, of course, rejection, the body's violent reaction to a transplanted organ. It was not a new problem, having first made itself known in 1492. While Columbus was facing a mutiny at sea on his way to the New World, Pope Innocent VIII lay dying in Rome. His personal physicians had given up, but a foreign surgeon held out a shred of hope, and in desperation Church officials agreed to let him try the radical treatment he proposed. The doctor, whose name, perhaps with good reason, was unrecorded, argued that an infusion of young blood—in effect an organ transplant—would rejuvenate the Holy Father and provide him with many more years of life.

Blood from three young boys was injected into the Pope's veins. The boys were bled to death. The Pope died too, and the imaginative but unsuccessful physician fled. But the idea survived, and Samuel Pepys recorded in his diary that animal blood transfusions were performed in 1665 by an Englishman named Richard Lower. In France, Jean Baptiste Denis, physician to Louis XIV, transfused the blood of a lamb into a young man. The patient reportedly survived, but others were not as fortunate, and blood transfusions were prohibited in France, England, and Italy for many years thereafter.

The advent of blood typing, at the beginning of the 20th century, made blood transfusions practical and also indicated a degree of genetic relationship that, while not completely accurate, was reasonably close.

When Marius's mother agreed to donate one of her kidneys to her son, Dr. Hamburger grew optimistic. "Our hopes were based on three factors: One, the boy was young and vigorous; two, donor and recipient were closely related. Finally, the red blood-cell groups of mother and son, which are an excellent indicator of just how close the genetic relationship is, were a close match for each other."

It was a valiant effort. The transplanted kidney functioned for twenty days. But on the twenty-first day after the operation, it was rejected. Marius died, but his case became a medical milestone.

"This was a longer survival," Dr. Hamburger wrote, "than in any previously published case, and it suggested that the close genetic relationship between recipient and donor might be important to survival of a grafted kidney."

Dr. Hamburger's supposition was swiftly confirmed when a kidney was transplanted from one identical twin to another and survived in its new host. The historic transplant was performed in Boston in 1954 and opened the door to hundreds of successful kidney transplants that have followed. But the same basic immunological reaction that rejected the life-saving kidney in Marius's body so many years ago is still the major barrier to the successful transplantation of all the organs vital to life.

Left to its own devices, the body will always reject foreign tissues, failing to distinguish between a beneficial organ composed of nonself tissue, and the most minuscule, innocuous bacterial invader. The reaction against each will be essentially the same, a savage, unceasing attack by the body's immunological defenses. It is this immune response that insures the individuality of the being and his biological integrity. Only thus has mammalian life been able to evolve and survive.

"The first point we need be sure about," declares Nobel Prize-winning Australian immunologist Macfarlane Burnet, "is that immunity is something that has had a significant effect on survival in the past—that it has been molded under specific evolutionary pressures and is not something that has emerged more or less accidentally in connection with some only distantly related bodily function or evolutionary requirement."

What is this marvelous and awesome mechanism that protects so fiercely, so blindly? The human body consists of trillions of cells and where defense is concerned each has a mind of its own, attuned to but one idea—survival. Let an invader seep through the protective envelope of skin, or enter by any other route, and a complex set of events begins within the body with the ultimate goal of repelling the invader. At the site of invasion, on the surface of the skin or in the mucous membranes that line the body's openings, tiny sprays of enzymes—the catalysts of body chemistry—start flash-fire reactions to tear down and dissolve foreign substances. From the blood, where it maintains a prowling vigil, comes a type of white blood cell known as a phagocyte to engulf and devour the invader. At the same time, other white cells and biochemical agents swirl into the area, damming it off from the rest of the body, so the primary defenses can complete the job of destruction. But if the invasion is massive enough, or virulent enough, the wall is breached and the invader rockets through the circulatory system, carrying infection to the farthest reaches of the body. It is at this point that slower-moving but more powerful defense forces are employed.

In the lymphatics—the watery network that meanders like Venetian canals through the tissues, cleaning, filtering, and disposing of waste products—a potent, explosive weapon is manufactured. This is a special protein molecule called an antibody, which might be described as a molecular guided missile, precisely designed to search out, lock onto, and destroy a particular invader called an antigen.

The exact relationship of the antibody to the antigen was the source of great controversy among biologists. There are two general explanations of the phenomenon. One of them, called the instructive theory, holds that specific antibodies are fashioned in intimate contact with an antigen. The antigen acts as a template that serves to instruct the antibody-generating leukocyte to design an exact fit. The effect is a sort of lock-and-key mechanism.

New research indicates that this in fact may be the mechanism involved in producing one form of antibodies—those responsible for combining with and deactivating disease-causing bacteria and viruses.

The antibodies that attack foreign tissues or transplant antigens, however, seem to have a somewhat different method of formation. Their birth is explained by the clonal-selection theory, which holds that the antibody-producing cells receive all the information needed to manufacture antibodies when they receive their genetic code. Thus, each different antigen acts merely as a fuse to ignite these specialized cells into explosive growth, so that whole clones, or colonies, of antibody-producing cells are produced.

Until recently, the two theories had engendered loud arguments between their respective supporters, but little in the way of experimental evidence. Now it appears that both theories are correct and that there are two different immunological systems; one associated with the circulating antibodies that are products of instruction, synthesized in the blood in response to disease-causing agents; and the second associated with these antibodies created by clonal selection as a response primarily to foreign-tissue antigens.

Experimental proof of clonal selection was recently offered by Dr. G. J. V. Nossal, director of the Hall Institute of Medical Research in Melbourne, Australia. "There is now *direct* experimental proof that antigens provoke antibody formation by stimulating proliferation of specific colonies of antibody-reproducing cells."

By tagging certain lymphocytes with radioactive iodine, the researchers were able to follow the trail of an antibody-forming cell after it was exposed to an antigen. They found that each antigen-sensitive cell launched itself upon a series of explosive divisions—eight or nine times—creating an entire colony of antibody-manufacturing cells where before the administration of the antigen there had been only a few.

Nossal's report lent credence to an idea many scientists held, that the cells which produce antibodies are themselves produced in a puzzling gland knows as the thymus, which is located behind and atop the breastbone, just below the neck. In the child the thymus is almost disproportionately large and divided into two oval-shaped lobes. In the first year of life the thymus triples its weight, and it keeps on growing up to the age of ten. Then the thymus falls behind and slowly shrinks. When the child has become an adult the thymus is barely distinguishable from the fat in which it is embedded.

Despite this unusual behavior, the thymus seems charged with one of the most important tasks in the body and must accomplish it early in life. The job is to stockpile cells called leukocytes. During the early stages of life, the thymus produces specialized leukocytes, called lymphocytes, at a prodigious rate and ejects them into the lymphatic system as freely wandering cells. Some set up reproduction centers in the regional lymph nodes—clusters of lymphoid tissue found in the armpits and groin and at other locations in the body—where, for the rest of life, they continue to divide and spill new lymphocytes into the lymphatic canals. Others settle down in the spleen and bone marrow, where their descendants are launched on a prowling course through the bloodstream. Throughout life, the leukocytes are charged with the body's defense, functioning both as bacteria-destroying white cells and as the founding fathers of virus-neutralizing antibodies. The specialized immunological ability involved in the clonal-selection response to foreign tissues is seemingly predetermined by the leukocyte-producing thymus.

Just how the thymus instructs the leukocytes in the performance of their varying duties remains unclear. The most promising explanation to date involves DNA. The most important component of the cell nucleus is, of course, DNA, and in mammals the thymus is the greatest producer of DNA. The leukocytes produced by the thymus contain in turn the greatest ratio of nucleus to surrounding cell matter of all animal cells. The high DNA content in the leukocytes is what presumably enables them to distinguish between self and nonself and also what orders them to make antibodies to neutralize or destroy any foreign invader.

Further understanding of the process of antibody formation will be of great value in preventing transplant rejection. For with the body's immunological defenses operating, the ultimate fate of most transplants seems to be rejection.

One exception is the cornea, the transparent coating that covers the front of the eye. It cannot be nourished directly by blood vessels, since they would block light from entering the eye. Thus, nature employs a different method, osmosis, which allows the nutrients to filter through the cellular walls of the cornea. But lymphocytes cannot penetrate the cornea by osmosis, and so rejection is not a problem with corneal transplants. With other tissues, however, rejection and its prevention is the major problem faced by transplant researchers.

The methods used to prevent rejection have been likened to using a cannon against a mosquito. Radiation prevents the formation of antibodies, red blood cells, and other essential blood components; drugs, primarily anticancer agents, are toxic and have a shotgun effect. One of the most effective and widely used drugs is called Immuran. Originally developed as an anticancer agent, it will blunt the immune response swiftly and effectively, like breaking the point off a knife. But it does so for all the antibodies present in the body. Most of the bacteria always present in the body are ordinarily harmless and kept in check by the white cells on their vigilant patrol or by latent immunity. Other, more powerful antigens require the antibody mechanism to tool up to moderate production to keep the body free of infection. But when Immuran is injected, the immunizing factors are paralyzed and under these circumstances even usually impotent bacteria can show the biochemical claws of the fiercest microbial predators. The individual is left defenseless in an environment suddenly fraught with menace. "We have our choice," said one transplant surgeon ironically. "We can get the transplant to take, by using enough drugs to prevent rejection, and then have our patient die of a common cold, or we can go easy on the drug and lose the graft to rejection and thus lose the patient for lack of the transplanted organ."

Most surgeons prefer the risk of infection, trying to tread a very

Dr. Paul Terasaki of the University of California at Los Angeles Medical Center maintains a tissue-typing service for many of the nation's transplant centers. Terasaki uses a computer to match the white blood cells of potential donors to recipients.

fine line between rejection and lowered immunity. One of the best hopes is to get a donor close enough genetically to the recipient so that a minimal amount of immunosuppressive agent will be required.

"I strongly suspect that most—if not all—of the patients who have reportedly survived for a reasonable length of time with kidney transplants from other than identical twins have done so because, quite by chance, they got kidneys from persons with matching tissue," says Dr. John P. Merrill, head of the renal laboratory at Boston's Peter Bent

Brigham Hospital and a member of the team that in 1954 performed the first successful kidney transplant in history.

Merrill thinks that in terms of the immune response every human being appears to be a sort of identical twin to about sixteen per cent of the population. Thus, in theory, any patient requiring a transplant should be able to draw on sixteen compatible donors from a random population of one hundred. But how to select those sixteen? The answer lies in an effective means of typing and matching tissue just as we now type blood. The problem is that tissue is far more complex in terms of compatibility. "In red blood cells, there are more than thirty antigens known," says Dr. Paul Terasaki of the University of California at Los Angeles, "but only a few are important in transfusions." No one yet knows just how many white-cell antigens are implicated in tissue rejections. "One of our major efforts," explains Dr. Terasaki, "is to find out how many there are in white blood cells and which are important."

Terasaki takes white-cell samples from the recipient and prospective donors and mixes each sample with a panel of lymphocytes that presumably represents the entire spectrum of tissue-antigen possibilities. The reaction of both donor and recipient cells is noted and the correlations are made by computer. The closeness of the genetic relationship is established with some degree of accuracy, though not as accurately as can be done in blood typing.

Still, the combination of tissue typing and a new drug that seems to confine its activity only to the antibodies responsible for rejection has resulted in some dramatic transplants at the University of Colorado. Here Dr. Thomas Starzl, a transplant pioneer, has been using an agent called antilymphocyte globulin (ALG) to prevent rejection.

The serum is produced by injecting a concentrate of human spleen, thymus, and lymph nodes into a horse. The horse's immunological system promptly manufactures antibodies to fight this distinctly foreign antigen. The horse's blood is then drawn off and refined until only the gamma-globulin fraction—that part of the blood which carries most of the antibodies—and a few other traces of other blood elements remain. This is the ALG, which is then injected into a patient who has received a transplant. The ALG seems to act as a decoy, getting the patient's immunological system to attack it instead of the transplant.

"It seems to send the body's lymphocytes off on a wild-goose chase," explains Dr. Starzl.

The results have been spectacular. ALG combined with tissue matching and small amounts of the more familiar antirejection drugs prevented rejection in twenty-seven of twenty-nine transplant patients. Since he began using ALG in 1966, Dr. Starzl had been able to report: "The mortality in our clinic in one year is going to be about five per cent as opposed to the thirty-five per cent that has prevailed here and everywhere else."

More efficient tissue-typing techniques and ALG's success encouraged Dr. Starzl to attempt what may be the most difficult of all transplants—the liver. Until 1967 only a handful of liver transplants had been attempted, most by Dr. Starzl, and survival had never been more than a month. But the need for a means of protecting the liver from rejection is vital, as the liver, unlike the kidneys and the lungs, is unpaired, and as yet no machine can take over its functions, even temporarily, as the artificial kidney or the heart-lung machine can for those organs. A healthy, functioning liver is absolutely essential to life. So important is it that nature has bestowed upon the five-lobed liver the ability to regenerate portions of itself, something no other organ can do. The liver makes bile, a fluid vital for digestion, and regulates the amount of glucose sugar in the blood by breaking down and storing any excess after a meal. It stores vitamin A and manufactures fibrinogen, a blood element needed for clotting, and heparin, an anticoagulant. Specialized cells in the liver, known as Kupffer cells, help to dispose of weak and used-up red blood cells and help fight against disease.

Beginning in July 1967 Dr. Starzl and his surgical team performed five liver transplants. The recipients were all infant girls, tissue-typed by Dr. Terasaki and given ALG. The purpose of the typing was to rule out donors so far removed genetically from the recipients that there was no hope for long-term retention. The typing indicated that one donor-recipient pair was strongly incompatible, three pairs were moderately incompatible, and one donor and recipient were classed as only slightly incompatible.

Two of the children died, though they survived longer than anyone who had received a liver transplant had ever lived before. One of the two who died had been classed as the poorest match with the donor; the other had a moderately incompatible match. As of this writing, the three other babies are still alive, their donated livers performing the complex biochemical functions so necessary for life.

As is so often the case, an interdisciplinary approach can often solve a problem that seems insoluble when viewed from the narrow focus of a single specialty. Thus the burgeoning medical field of fetology may have a solution to the immunological problem of rejection of surgical transplants. The answer may lie in that most wondrous of organs—the placenta.

Before the birth, and for some months thereafter, the baby seemingly has little or no working immunity. The placenta acts as a very effective immunological filter, screening out all disease-carrying bacteria and most viruses. At the same time it permits passage from the mother to the fetus of antibodies that will protect the fetus while still in the womb, and the baby for some months after birth from any disease to which the mother is immune.

If during this period of immunologic incompetency a foreign antigen—any protein that will provoke an antibody response—were injected into the newborn or fetus, the body might simply note its presence and forever after regard it as familiar or self. Then, when the immune mechanism is later turned on, all future exposures to the same material will not trigger an antibody reaction and subsequent rejection.

This at least was the theory propounded by several prominent immunologists a few years ago. Recently, another immunologist, Dr. Bernard Pirofsky, of the Oregon University School of Medicine, tested the thesis, first on animals and then on human fetuses. Working with a research team in Mexico City, sixteen pregnant women carrying fetuses with type O blood allowed their unborn children to be exposed to type A blood antigens. The material was simply injected with a needle through the mothers' abdomens, through the placenta, and into the amniotic fluid. The fetuses, all within three to four weeks of term, were in no danger and simply swallowed the antigen as a normal part of their habitual sipping of the amniotic fluid. Months later, after the newborns developed their immunological competence, their type O blood was exposed to type A blood. There were two reactions possible. If the type A antigen was recognized as foreign, the two samples of blood, when mixed on a slide, would clump up, a sign of rejection. If the type A was to be familiar, that is compatible, with the O, an event that does not occur normally, the blood would not clump but would mix into a homogeneous pool. The blood did not clump in a single case. The type A blood was not rejected by the type O babies' blood. Rejection had been forestalled.

The implications are astonishing, for it is now considered possible, in terms of tissue at least, to make each person the identical twin of every other human being and thereby a potential rejection-free donor to and recipient from everyone else. To Dr. Pirofsky, the idea is scientifically valid.

"If," he notes, "we can obtain groups of antigens, the materials that make people different, and we know just where they are currently and how to isolate them and obtain them in sufficient quantities, it could be possible to expose children at the time of birth, or before they are born, to all the differences of human beings. And if this can be preserved, when that individual grows up, all human tissue would be part of his own body. If a severe burn should occur, a transplant of skin could be given. If a heart defect occurs, any other human heart could be transplantable."

It seems that, well before the 21st century begins, the immunological barriers to transplants will have been hurdled, but other problems remain to be solved before this startling form of spare-parts surgery can become a clinical routine. Chief among them is the question of supply. Where are all the new, healthy kidneys, hearts, lungs, livers, and other organs going to come from? Only the kidney can be donated by a living human volunteer with any degree of safety for the donor.

The most likely source of supply now and in the 21st century would appear to be the dead. Everyone must die—some of disease, some of old age, and some accidentally in the first flush of youth. When death comes and respiration ceases, when the heart stops and the brain ceases to flash its signals throughout the body, some elements of the body—tissues, bones, even organs—cling to their own lives a bit longer. Jolts of electricity can briefly restore the beat to a stopped heart. Vital organs can be flooded with a chemical solution to prevent the blood from clotting in the vessels and to slow the process of dying. If rushed into a new host and supplied with blood, these organs—barring infection and ultimate rejection—can function indefinitely in their new environments. These facts make the prolongation of life possible through the use of cadavers.

The use of cadaver parts is not a new concept. The transplantation of corneas from the dead to the living has been an accepted practice for more than fifty years. The precedent is established. Cadaver kidneys now provide the major source of supply for the hundred or so kidney grafts done each year. The spate of heart transplants has been made possible only by the use of cadaver hearts, snatched hurriedly from the dead and just as hastily implanted into the dying in the hope that life will be the end product.

The welter of moral and legal problems entwined in the taking of a cadaver organ is formidable. Some states make it difficult, or even impossible, to gain permission to take an organ soon enough after a person's death to make a transplant possible.

Bereaved relatives, the next of kin who must give their consent virtually the instant after they have learned of the death of a loved one, must undergo a terrible emotional trauma when asked. In the circumstances, their answer is as likely to be no as yes.

Will we, in the 21st century, have the right to consign perfectly good organs to the grave, or shall the body be stripped of its still-functioning organs the moment death is at hand?

With this question raised, there comes the even knottier one of just what is death, when precisely does it occur? "For the person who takes a vital organ too soon," one surgeon declared, "society has a word—and that word is 'murder.'"

"We're going to have to have a new definition of death," says Cleveland neurosurgeon Dr. Robert White. White, the only man in history completely to isolate the living brain (he has also transplanted these brains, from one monkey to another, and kept them alive for periods of up to thirty hours), hopes to understand brain function better by these experiments. "The old medical idea that when the heart and lungs stop, a patient is dead, of course, will no longer hold; it doesn't hold today. The essential organ system we have to think about for death is brain, and what we must have defined for us is a period, or really if not a period at least an area where we can be absolutely sure that the individual will not recover even though his heart and lungs are working. There are many patients in hospitals who are in a coma,

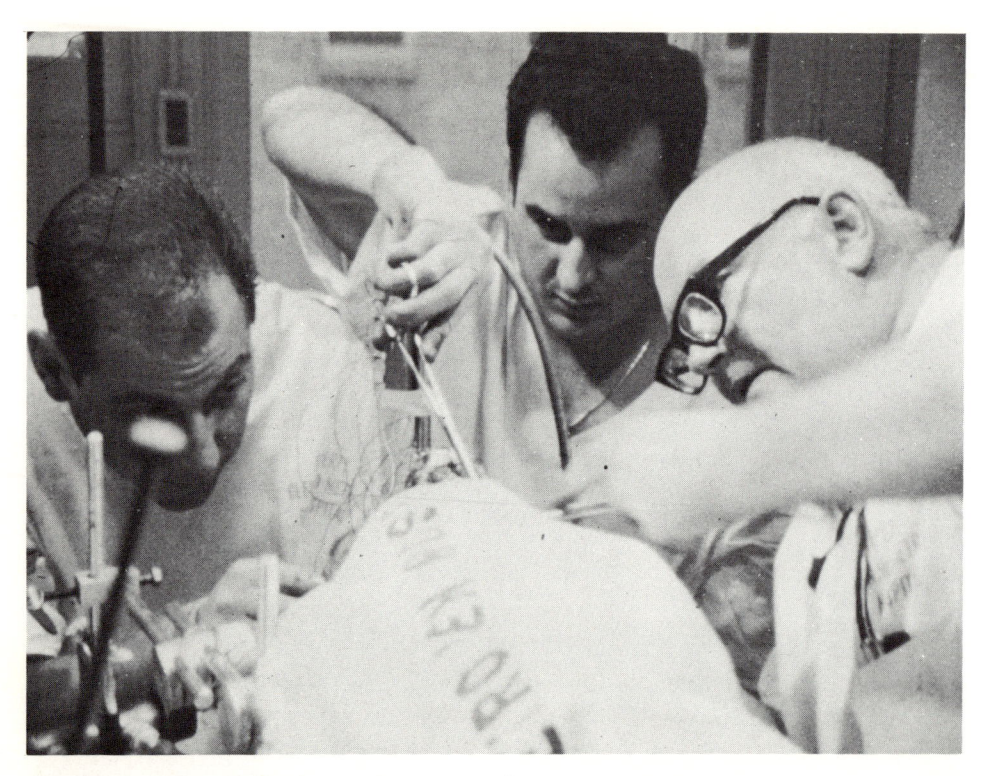

Dr. Robert White (at right) and his research team at Western Reserve University Medical School detach the body of a rhesus monkey after first isolating its brain.

and they have fine heart functions, they have fine lung functions, and so forth. But these individuals are destined never to recover and probably will die in hours, days, weeks, or, unfortunately, in months.

"Now the problem here is if we knew by some chemical test, by some electrical test, or perhaps by some examination that these individuals are never going to recover, then the question would be upon us whether or not we should define these people as in a category of nonrecovery and as a consequence be justified in removing one or more of their organs."

When this chemical or electrical test Dr. White speaks of has been devised, and many feel that a completely flat brain wave on an electroencephalograph for a period of five minutes is that test, what

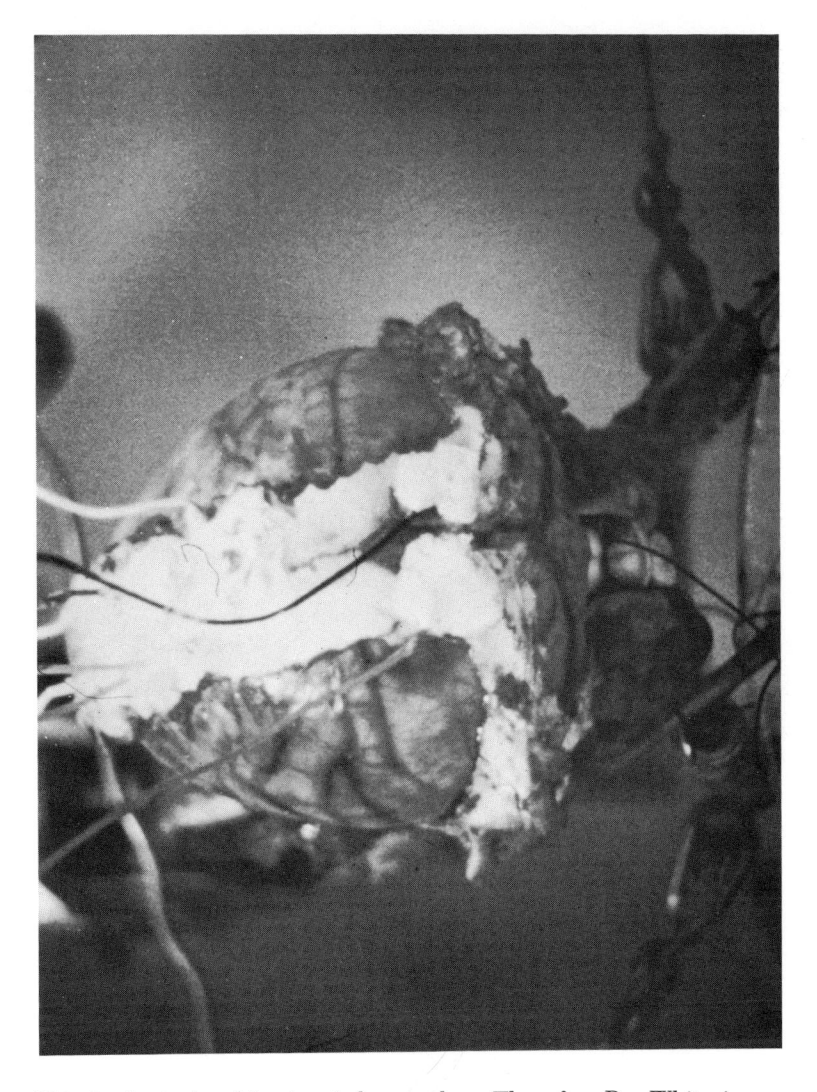

This is the isolated brain of the monkey. Thus far, Dr. White is the only scientist to successfully isolate and maintain the brain of a primate for any length of time.

then? Will the surgeons be forced into a macabre death watch, waiting to snatch organs from the just-dead? Will they hurry their patients into incredibly difficult surgery simply because the organ is suddenly available, even though the patient may not yet be prepared to survive the shock of surgery?

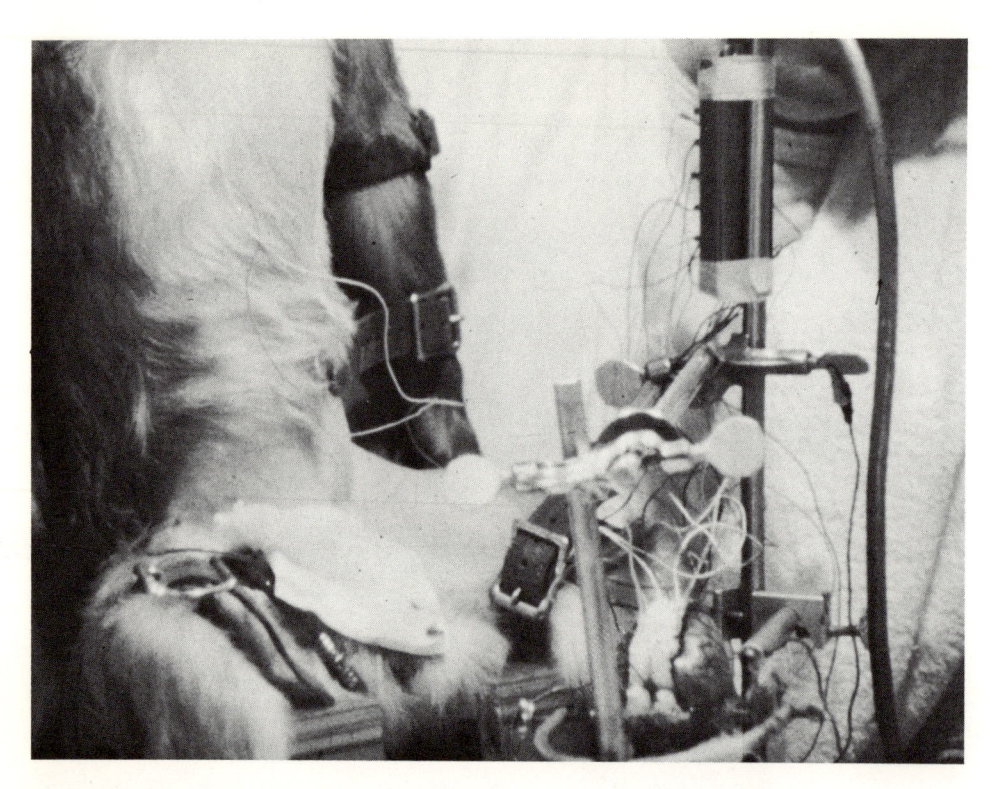

The brain (lower right) is transplanted into the body of another monkey to keep the isolated brain alive so that intensive studies never before possible can be made. The recipient monkey provides kidneys to cleanse the blood that nourishes the isolated brain.
Such brains have been kept alive for approximately thirty hours.

The obvious solution, from a moral and a medical standpoint, is the establishment of an organ bank, where cadaver parts will be stored until they are needed. Blood banks first established in 1937 at the Leningrad Institute of Blood Transfusion set the precedent and were followed less than ten years later by eye banks. The first of these storage vaults for corneal bequests was set up in New York City in 1945 as the Eye Bank for Sight Restoration, which provides more than four hundred corneas for transplantation each year.

But the problem of storing whole organs is more complex than that of storing blood or corneas. The most promising technique at this point appears to be the deep freeze, but there are enormous problems. At any temperature between 32 degrees Fahrenheit and 202 degrees below zero Fahrenheit, ice crystals form in and around cells, shredding their gossamer membranous walls. Below —202 degrees, however, an extraordinary event takes place. The water in and around cell walls turns to smooth, glassy ice that simply encases cell walls without disrupting their delicate chemical equilibrium or tearing the walls.

This was the promise offered by cryobiology, a field of science resulting from the marriage of cryogenics—the physics of extremely low temperatures—and biology. The temperature scale in this relatively new technology resembles a degree of cold below that found on the moon during lunar night—down to the bottom of the scale at about 450 degrees below zero Fahrenheit. In this strange world, fantasy replaces the normal realities of the physical world. Air freezes rock-hard. Some metals become as brittle as glass, and electrical resistance drops to zero. Most life processes come to a stunning halt, but without the finality of death. When rushed back to normal temperatures, they pick up the threads of existence as if nothing had happened.

The first organ to go into the deep freeze of cryobiology was blood. In 1949 Dr. Basile J. Luyet of the Rockefeller Institute discovered that seventy per cent of the red cells in ox blood survived the very low temperatures, provided the freezing was accomplished rapidly enough to prevent the cell-damaging crystals from forming. Using liquid nitrogen, Dr. Luyet was able to lower the temperature of human blood by two hundred degrees a second. The resulting glass-smooth ice did not damage the cells.

By 1960 techniques were worked out that would permit the rapid freezing of blood, so that it could be stored for years instead of the usual twenty-one days. One process used a rocker arm that agitated a container of blood while it dipped it in liquid nitrogen. In forty-five seconds the blood was dropped from room temperature to 230 degrees below zero Fahrenheit. Tests revealed that the damage to the cells was minimal—eighty to eighty-five per cent survived the rapid freeze.

But another problem to be overcome was thawing, since ice crystals form as readily when the temperature climbs upward through the critical zone as when it drops below it. Containers of blood were immersed in warm water and gently shaken. Stone-hard blood liquefied in seconds with this technique.

The obstacles to freezing whole organs appear to be more formidable than those encountered with blood. Damage caused by the formation of ice crystals is not the only problem. When tissues are frozen, the electrochemical equilibrium that makes the biological engine go is disrupted. Electrolyte solutions, vital to life, become concentrated when frozen; the delicate balance seemingly cannot be restored when thawed. It may not be until the 21st century that cadavers will be mined and their still-healthy organs deep frozen in wholesale lots, stored indefinitely until needed.

But some progress is being made today. A well-known commercial solvent, dimethylsulfoxide, or DMSO, which had a brief flurry of fame when some researchers likened it to a miracle drug, is being used as a sort of antifreeze to protect organs from the destructive effects of freezing. Dr. J. H. Bloch, at the University of Minnesota Medical School, has preserved animal organs for as long as two weeks. At the Protein Foundation in Boston, bone marrow has been successfully frozen and preserved for as long as two years, simply by adding DMSO.

An alternative freezing method being worked on in Minnesota by Dr. Richard Lillehei may have immediate practical applications. Dog kidneys, which have a forty-five-minute life without the normal blood supply or an artificial one called perfusion, and which had never been stored successfully for more than six hours, have been kept viable for twenty-four hours in tests that involve not only freezing but also increased pressures similar to those found below sea level. Placed in a tank that is in reality a giant pressure cooker, the kidney is perfused all the while the temperature is being lowered. In this state of advanced chill rather than freezing, it has been kept for one or two days and then reanimated. Reimplanted in the dog, the stored kidneys performed quite satisfactorily.

For long-term storage, Dr. Lillehei expects his method to be essentially the same, but with the temperature dropped well below freezing, into the cryogenic areas. As yet, however, the major problem remains thawing.

Despite the difficulties, an organ bank has been started by seven hospitals in the Los Angeles area. Called the Los Angeles Transplantation Society, the bank takes organs, with the permission of relatives, from the bodies of people who die in the hospitals. The organs are available to recipients in any of the hospitals. The bank was made possible by the recent development of a perfusion pump that will keep the kidney nourished and supplied with oxygen for three days. Without this pump, surgeons faced a six-hour deadline in removing the kidney and getting it into the recipient before the tissue died. Now organ storage for three days at least is a reality, and organ banking even on such a short-term basis becomes possible.

The need for living spare parts will increase in the 21st century as the skills and techniques of the transplant surgeons increase. Banks of cadaver organs may not be able to keep up with the demand. But what of a living organ bank? Many experts have suggested the breeding of animals specifically as a source of organs for transplants to humans.

Desperation transplants of animal parts have already been made by surgeons. Five years ago at the Tulane University Medical Center a team of surgeons headed by Dr. Keith Reemstma transplanted the kidneys of a rhesus monkey into the body of a thirty-three-year-old housewife whose own kidneys had utterly failed. The monkey kidneys worked, not well, but worked nonetheless, for ten days before the surgeons decided to remove them. Placed on an artificial kidney machine, the woman died two days later.

A few weeks later, when no suitable human tissue could be found, the kidneys from a chimpanzee were transplanted by the same team into a forty-three-year-old man in the terminal stages of chronic kidney disease. Massive doses of radiation and drugs blunted the rejection response, but at a terrible price. After two months, with virtually no immunological defenses, the patient died of pneumonia.

It is evident that the antigenic gulf between man and monkey is wide, but perhaps it is not unbridgeable. Biochemical comparisons of blood and proteins manufactured by both man and ape have established a rather close relationship. Ape and man might even be considered to be virtual blood brothers. Two other members of the primate family, chimpanzees and baboons, offer man a complete range of blood types. Baboons have groups A, B, and AB, but no O. Chimps have A and O, but no B or AB.

This close blood tie has led Dr. Morris Goodman, associate professor of microbiology at Wayne State University in Detroit, to declare, "Some existing apes have more in common, ancestrally speaking, with man than they have with other living apes."

Man's ancestral links with other animals may also provide the means of insuring a supply of organs that will be completely adequate to the demands of transplantation in the 21st century. "I foresee the day," said Dr. Claude R. Hitchcock, professor of surgery at the University of Minnesota, "when animal ranches, providing a constant supply of primate organs, will be established for transplantation purposes."

To bridge the gap that exists between man and animal, Nobel Prize-winning geneticist Joshua Lederberg of Stanford University offers this possible solution.

"A vigorous eugenic program, not on man, but on some nonhuman species, might produce genetically homogeneous material as sources for spare parts. The technical problem of overcoming the immune barrier would be immensely simplified if the heterografts came from a genetically constant source, the more so if the animal supplying the grafts could be purposely bred for this utility. At present, the only adequately inbred mammals are small rodents."

Lederberg has a 21st-century solution that is already being acted upon to some extent at Georgetown University. Here Dr. Charles Hufnagle, a pioneer developer of artificial heart valves, is attempting to treat unborn calves in such a way as to render them genetically neutral. The aim is to provide a genetically compatible, anatomically similar heart for transplant purposes.

If the effort is successful and humans accept calves' hearts, a living bank could be established.

"Such a bank," explains Dr. Hufnagle, "would avoid the moral and ethical problems involved with waiting for a human being with a healthy heart to die and then taking his heart. Also, a critically ill potential transplant recipient would not have to wait at the brink of death for a suitable donor to appear."

This is precisely the situation today, one that may become intolerable in the 21st century. But with such a bank, transplantation of almost every organ, even the one considered the font of life, the heart, would, in the 21st century, reduce miracle to routine.

6 | The Human Heart

On December 3, 1967, at the Groote Schuur Hospital in Cape Town, South Africa, Dr. Christiaan Barnard and a thirty-man surgical team transplanted the living heart of a twenty-five-year-old girl, killed by a car, into the chest of fifty-five-year-old Louis Washkansky. The transplant was a success, the first in history, for Washkansky lived and grew stronger. Then, eighteen days after the operation, Louis Washkansky died of pneumonia, but the transplanted heart itself continued to beat to the moment of death.

"What we needed," said Dr. C. Walton Lillehei, chief of surgery at New York's Cornell Medical Center and himself a pioneering heart surgeon, "was a Moses, someone to lead the way. Chris was that leader, and I think we'll see more heart transplants done as a result." Dr. Lillehei was participating in a round-table discussion on December 27, only twenty-four days after that first heart transplant. A year later, one hundred heart transplants had been attempted and forty-three of the patients were still alive, the longest survivor being Dr. Philip Blaiberg, another of Dr. Barnard's patients, who received his new heart in January 1968.

Yet heart transplants, dramatic though they be, do not come close to solving the truly gigantic problems of heart disease. Each day seven thousand Americans suffer heart attacks. Every year half a million

Dr. Christiaan Barnard, who performed the world's first human-heart transplant in Capetown, South Africa.

people die of heart attacks in the United States, while the total of deaths due to all forms of cardiovascular disease is at least one million persons a year, more than all other causes of death combined.

However, a heart attack need not mean instant death. Coronary-care units are enabling more and more people to survive the initial attack and let nature heal the scarred and weakened heart. New drugs, new surgical procedures, and perhaps most important of all, a profound, though not very remarkable, change in the American way of life may be the most significant factor in conquering heart disease. A tendency in America toward dietary changes and increased exercise may play as great a role in preventing heart disease as any new breakthrough in medicine. Even where surgery is indicated, the most dramatic of operative procedures, the heart transplant, will not become a commonplace in the 21st century.

"I think," said Dr. Barnard, "that there are certain heart conditions which we'll still treat in the way we treat them today. They are a very minor sort of heart ailment, which can be very adequately treated. I think, for example, of one condition, a hole between the upper chambers

Dr. Philip Blaiberg, the world's second heart transplant patient and the longest-lived, shown here leaving Grote Schuur Hospital after surgery.

Dr. C. Walton Lillehei, one of the world's foremost cardiovascular surgeons. Now chief of surgery at New York Hospital, Dr. Lillehei recently made history by transplanting the heart, kidneys, and pancreas from one donor into four patients.

of the heart. This can be sutured and closed very adequately and the patient has a normal heart afterward. I don't think anybody would dream of replacing a heart in such a person. I think that, even in the 21st century, this form of treatment [the transplant], even if it becomes very routine treatment, will still be applicable only to certain hearts, the more severe heart conditions."

In a figurative sense, it seems irrational, almost unnatural, for the human heart to fall victim to disease. Nature lavished such skill and care on its construction and design that it seems a pity, almost an insult, for the heart to go bad.

The human heart is an elegant complex of muscle and tissue, exquisitely engineered to perform a simple but absolutely vital function. The heart is a pump, charged with a burden of circulating the equivalent of 4300 gallons of blood through the body, day after day, year after year, in beats that are the tempo of life itself. If the beat stops for five to fifteen seconds, the body staggers to a fainting halt. A slightly longer pause and the supply of oxygen is pinched off from the brain and convulsions may result. A longer period of oxygen deprivation—three to five minutes—and death is almost certain.

To prevent this, the heart is fashioned of powerful muscles, capable of contracting to one-half their maximum length. It is this great contractibility that provides the propulsive power of the heart. A network of nerves and sensitive chemical receptors govern the beat. The muscle fibers themselves have a sort of back-up system built in, in the form

A cutaway model of the human heart shows how blood is pumped through the lungs, where oxygen and carbon dioxide are exchanged. (*American Heart Association*)

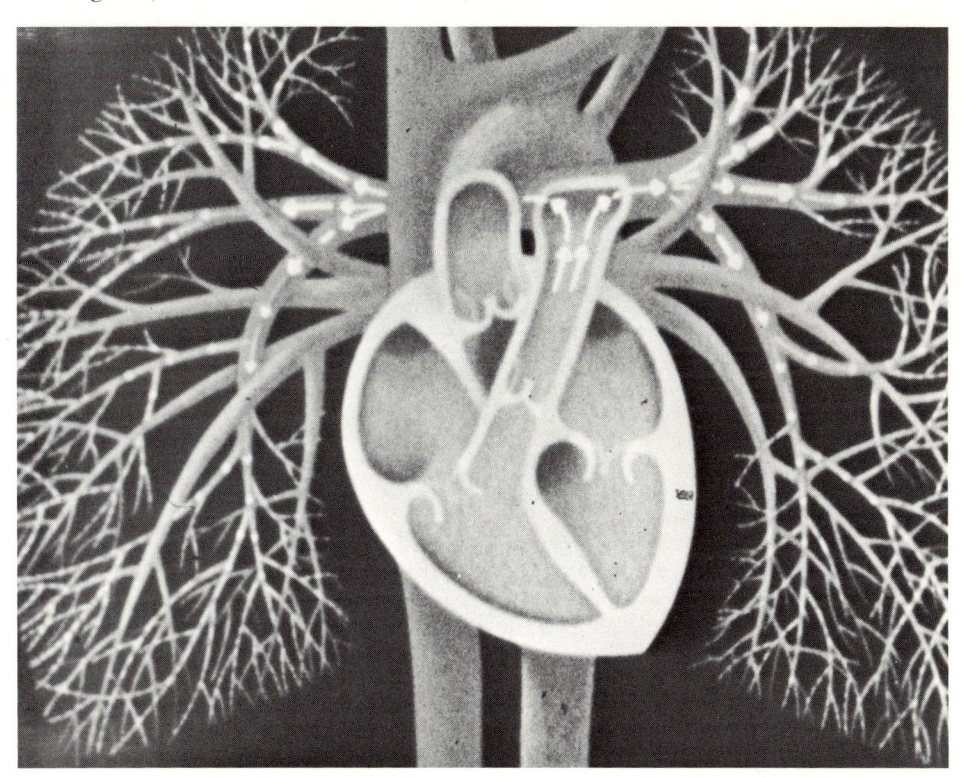

of syncytium—each connects with the other without any intervening cell walls. This direct contact allows each fiber to excite the next one into the life-sustaining contractions some seventy times a minute.

Although the regulating forces are complex, the anatomy of the heart is relatively simple. The adult male heart is about the size of a pair of clenched fists. It is composed almost exclusively of elastic muscle fibers nourished by blood vessels and activated by nerves and the principles of fluid dynamics.

Within its hollow walls of muscle, the heart is divided into four chambers, a right and left atrium and a right and left ventricle. The four chambers are perhaps best thought of as being two separate pumps, and doctors often class them as such—right heart and left heart—each containing an atrium and a ventricle. A simple system of valves keeps the blood moving in the right direction. Branching off the heart are the great vessels, the major conduits of the circulatory system which funnel the blood to the heart and from it through a vast network of branches and tributaries to all the tissues of the body.

Blood, drained of its oxygen by the cells of the body, enters the venous return system and is channeled into the right atrium. As the chamber fills up, it contracts and, developing a pressure differential, forces the blood through a valve and into the right ventricle.

Now the ventricle contracts and drives the oxygen-starved blood through the pulmonary arteries into the lungs, where the gas exchange—carbon dioxide for oxygen—takes place. The contraction is powerful enough to keep the blood flowing on through the pulmonary vein and into the left atrium.

The process of contraction then forces the oxygen-rich blood from the left atrium into the left ventricle and from there into the great trunk of the aorta, whence it will go to all parts of the body, releasing its oxygen to the individual cells so that the entire organism may live.

The same circulatory system that nourishes the body must also nourish the heart. Without an adequate supply of blood the heart too will falter and die. Thus it is that the majority of heart attacks involve the blockade, or occlusion, of the coronary arteries, the vessels that feed the heart itself. A major contributor to these blockades is a chemical called cholesterol. Often thought of as a fat, cholesterol is actually a waxy alcohol manufactured primarily in the liver. Its main function seems to be to furnish an ingredient used by the body in the manufacture of some hormones, but even this attribute is not absolutely certain. "We still do not know of any absolutely necessary function that cholesterol in plasma performs," declares Dr. Meyer Friedman of the Mount Zion Hospital Medical Center in San Francisco.

There seems little question any more, however, about the role played by cholesterol in the development of atherosclerosis, the deposit of fatty layers on the inner wall of a blood vessel. These fatty layers build up as the cholesterol level grows, until they reduce the "bore" of the coronary artery to the point where little or no blood gets through. The result is the death of part of the heart tissue and a heart attack. If only a small proportion of the total heart muscle is affected, and the victim survives the attack, his heart will continue to maintain life, but at a reduced rate, as the once completely elastic muscle now has a section that is scarred and useless. After a period of time, the body generates new branches of the coronary arteries, a so-called collateral circulation, to increase the supply of blood to the heart. Thus, the victim of a mild heart attack can often have virtually one hundred per cent heart function restored by the body's own marvelous regenerative powers. However, additional blocks may form and other heart attacks, possibly fatal, may occur.

One of the most imaginative approaches to this problem was made by Dr. Arthur M. Vinberg of the University of Montreal. Vinberg stripped a branch of the subclavical artery, which feeds the chest, and plugged it into the heart, in the same area that was receiving a diminished blood supply from one of the coronary arteries.

The Vinberg implant, as it is known, has had a modest success. It shunts blood from the chest to the heart in about thirty or forty per cent of the cases. When it does work, it still takes up to thirty weeks after surgery before it begins to feed blood to the heart. In addition, the Vinberg procedure seems to be most effective on just one of the two main coronary arteries. Still, the idea is basically sound, and many lives have been saved with this procedure. It has also provided a great challenge to a number of cardiovascular surgeons to modify and improve upon it.

The best approach would be to plug the subclavical artery into the blocked coronary artery itself at a point below the block. Surgeons have known for a quarter of a century now that in almost seventy per cent of all cases of coronary artery obstruction, the block is limited to the first two to two and one-half inches from the origin of the blood vessel. But this approach would require incredibly precise maneuvering in an incredibly small area. The inner walls of the coronary arteries have a diameter of from one eighth to three sixteenths of an inch. Although slitting such a small artery open and then inserting the end of another one to make a tight, leak-proof connection seemed beyond the skills and technical capability of the surgeon, all the techniques deemed necessary did exist. Although ordinary operating microscopes provided enough magnification to view the coronary walls, they could not present enough of the operating field to the surgeon. Moreover, the placement of sutures in so small an area required a delicacy that even the most sure-fingered surgeons could not call upon.

Both of these major problems were solved recently by Dr. Charles Bailey and Dr. Teruo Hirose of the St. Barnabas Hospital for Chronic Diseases in New York. The two surgeons solved the vision problem by adapting a jeweler's loupe, which offered a 2.5 X magnification and at the same time kept the entire operative field within view, no matter how often the heart beat during the surgery. The small-suture problem was solved by slipping a metal ring over the end of the subclavical artery and then rolling the end of the artery back over the ring to make a cuff. The cuff was then inserted into the coronary incision and just one or two stitches were needed to hold it tight and leak-proof in its new site.

The first two cases in which this new technique was used were eminently successful. The surgery took only about twenty minutes, and, even more startling, both patients were out of the hospital in two weeks and back at work a month after the operation. X-ray motion pictures of one patient's heart a few months later showed the new coronary shunt delivering adequate supplies of blood to the heart, while above the junction the coronary artery was completely occluded. Without the surgery, the patient would unquestionably have died.

In surgery, as in most human endeavors, the most direct approach is often the best, and the most direct approach to blocked coronary arteries would be to slit them open and ream out the fatty plaques that block them off. Unfortunately, the procedure has proved extremely difficult in the narrow coronaries. Now a new method of reaming out blocked coronary arteries has been developed by Dr. Philip Sawyer, Dr. Martin Kaplitt, and Dr. Sol Sobel of New York's Downstate Medical Center.

The surgeons make a small nick in the blocked artery and then insert a hypodermic needle. Carbon dioxide under pressure is fired through the needle into the coronary artery, blasting the fatty plug loose from the arterial walls. Next, the surgeons insert a small forceps and tease the fatty core out of the artery. Plugs as long as four and one-half inches have been removed in this fashion, preventing what might have been fatal heart attacks.

Still another artery-blocking mechanism is the blood clot. Coronary thrombosis, or clots that form in the coronary arteries, may have the

A new method of clearing out a blocked coronary artery is to blast it open with jets of carbon dioxide under pressure. The gas is fired into the artery through a hypodermic needle.

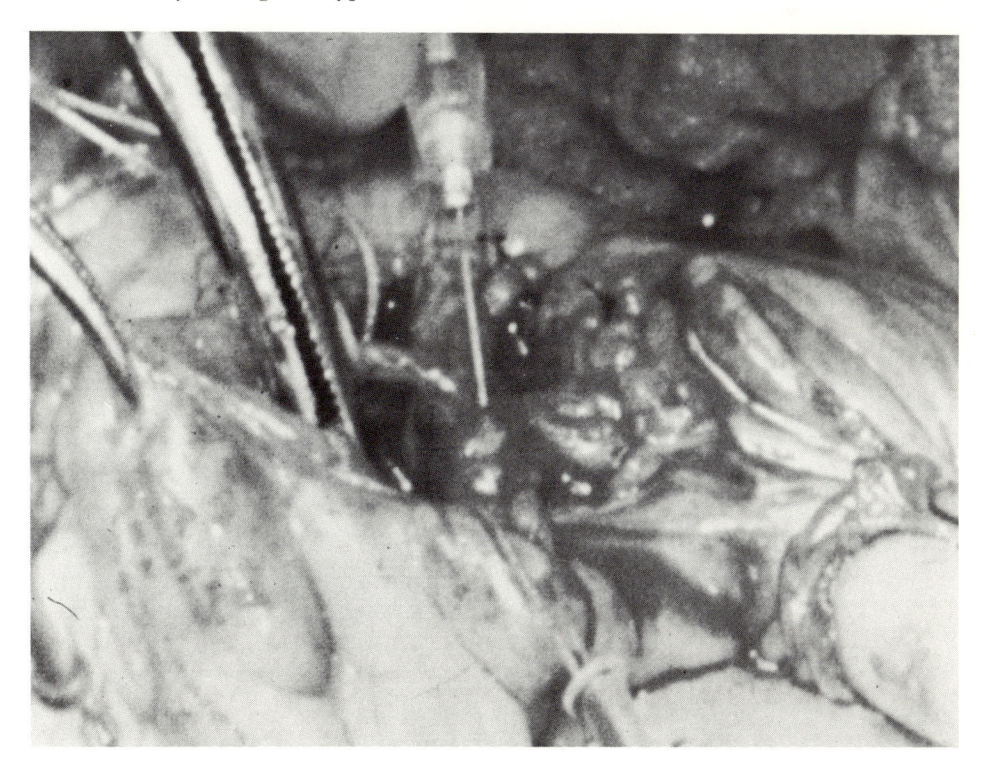

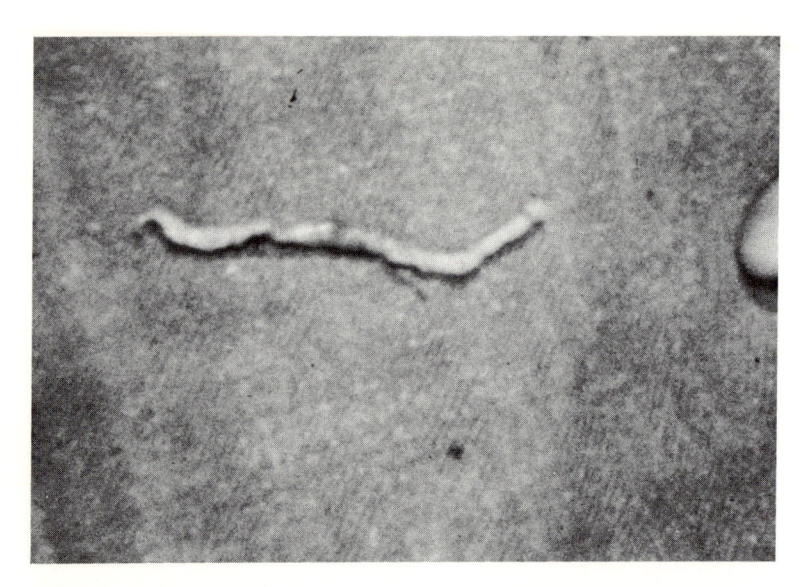

The fatty plug teased from a coronary artery after it has been blasted loose from the arterial wall. The technique was developed by a team headed by Dr. Philip Sawyer of the Downstate Medical Center in New York.

same obstructive effect as atherosclerosis. Surgery to clear blood clots from vessels is extremely difficult and dangerous and often impossible. Another alternative is to hope the clot dissolves spontaneously. Neither approach is considered satisfactory, and so a massive research hunt is on for a drug or drugs that will dissolve clots swiftly. The major problem is that the drugs uncovered so far not only will dissolve the clot but can also lead to uncontrollable hemorrhage. One recently developed agent, an enzyme called urokinase, shows promise of having a rather specific action just on the clot. Extracted from human urine, the drug does not in itself dissolve the clots, but rather switches on the body's own clot-dissolving mechanism. Undergoing clinical trials at Johns Hopkins in Baltimore, urokinase has been effective within twenty-four hours in about half the cases tested.

Still another drug approach is being developed by Laszlo Lorand, professor of biochemistry at Northwestern University. The Lorand method is essentially simple and involves "fooling" certain blood components into forming clots that are softer than usual.

A blood clot consists mostly of formed elements of the blood ensnared in a network of protein fiber. This protein is called fibrin and is normally absent from the circulating blood. Fibrin is formed when fibrinogen, a soluble protein in the blood, comes in contact with the air, or when a blood vessel is damaged.

Under the microscope fibrinogen is a long, rodlike molecule with rounded tips. So long as the tips remain, the individual fibrinogen molecules will not stick to each other and form the fibrin mesh that is the key component of a blood clot. The clotting mechanism—that is, the conversion of fibrinogen to fibrin—is triggered by an enzyme known as thrombin. Scientists have known for some time of the long and complex process that produces thrombin when it is needed, and occasionally when it is not, but the actual mechanism by which thrombin causes the creation of fibrin has been a mystery.

Lorand's research showed that, in the early stages of clotting, thrombin snaps off the rounded tips of the fibrinogen molecules. Without the tips, fibrinogen becomes the sticky protein fibrin, which adheres to other fibrin molecules like self-sealing envelope flaps. The result is a gelatinous substance considerably softer than the final clot.

Finally the clot becomes a tough, rubbery mass formed by crosslinking, a process in which the soft fibrin chains are cemented together in concrete-like slabs. This cross-linking was the second mystery Lorand solved. He found it involved a blood component that he named the Fibrin Stabilizing Factor, or FSF.

With these two clues firmly in mind, Lorand searched for a means of modifying the clotting process. He found a broad group of chemicals called coagulation inhibitors that are similar in structure to fibrin itself—so similar they "fool" other fibrin molecules. Thus fibrin joins the inhibitor in the same manner it joins with other fibrin molecules, but with one important difference. The final toughening process in clot formation, cross-linking, does not occur.

In its stead is a soft clot, strong enough to seal an open wound but weak enough to be dissolved gradually by normal blood enzymes.

The trapping of fibrin is a new and exciting approach to the problem of thrombosis, but it is still in the test-tube stage. But perhaps the pill to prevent coronary thrombosis may be in the 21st-century medicine chest.

An even farther-out drug approach to the problem of heart attacks is being taken by a pair of New Orleans scientists, Dr. M. Bert Myers and George Cherry of the Touro Research Institute. Myers, a surgeon, and Cherry, a physiologist, inadvertently discovered what appears to be a network of tiny blood vessels in the heart and other organs, which "come alive" only after death. The original direction of their research was aimed at solving a totally different problem. After massive surgery, usually for cancer, skin flaps when sutured down often died and were sloughed off. Textbooks had attributed this to sutures that were too tight. But Myers doubted this explanation, for the incidence of skin sloughs after surgery was virtually as high even when the sutures were loosely placed.

By raising long, rectangular flaps of skin in rabbits and then suturing them back again, Myers and Cherry found that the sloughing occurred only when the blood supply to the flap had been destroyed. The tension on the skin from the sutures had nothing to do with the life or death of the skin flap.

To see how much of the blood supply could be destroyed without killing the tissue, the two scientists injected a dye into the skin flap before it had sloughed off and followed its course on X rays through the blood vessels. The length of skin the dye traveled should then have determined the amount of circulation that remained to the flap. But to the amazement of Myers and Cherry, the area of skin that actually survived was about an inch longer than was indicated by the dye. Since more skin lived than had been predicted by the dye, it meant that some increase in circulation had occurred after the dye was given but before the cells that made up the tissue had all died.

"It seemed virtually certain," explained Dr. Myers, "that small collateral vessels exist which do not function during life, but which open up only after death."

After tests on rats, rabbits, dog, and pigs, they found that what was true for the skin was also true of the heart.

"What this means," declared the American Heart Association, "is that medical men eventually may be able to put this vessel system to work helping to stave off fatal heart attacks. Hopefully, activation of these new 'collateral' channels, by providing detours around the coronary arteries narrowed or blocked by disease, could help an oxygen-starved heart muscle receive an adequate supply of blood to lessen the extent of heart damage."

The next step was to find a drug that might open these channels before death. After a great deal of testing, Myers and Cherry found a family of drugs, called alpha-adrenergic blocking agents, that showed promise. These drugs work directly on the blood-vessel walls to counter the constricting effects of "stress" hormones, such as adrenalin. One of the drugs, phenoxybenzamine, proved to be quite successful, increasing the amount of tissue survival by sixty-five per cent after all of the visible major vessels to a flap of skin had been destroyed.

In dog hearts the drug has proved to be almost as effective, and the next step will be to test it on pig hearts, which anatomically are almost identical to human hearts. No experiments are planned for human beings for a while yet, but the experimenters feel the same principles should apply. Then a drug that might serve as an immediate first-aid measure for a heart attack might become a reality.

Still another first-aid device is under development at the University of Pennsylvania. Called a ventricular sleeve, it consists of a glass cup, lined with a plastic diaphragm that slips over the heart and squeezes it, pumping blood into the circulation. A compressed-air pump provides the motive force and the action itself is much like a milking machine working in reverse. The milking machine uses suction to draw out fluid while the ventricular sleeve uses the compressed air that is pumped between the glass cup and the plastic diaphragm to force blood out of the heart.

The sleeve has already been used clinically on six patients, all of whom had "died." In each case their hearts had stopped beating and could not be revived by any other means.

One of the patients, a forty-six-year-old woman, was maintained by the assist pump for over an hour, and then her own heart had rested and recovered enough to pick up the full burden of circulation. The other patients—most over seventy years of age—were kept alive for as long as seven hours. They were too ill, however, to recover their natural heart function and died after the assist device was removed. "Nonetheless," Dr. William Blakemore, head of the team that developed the ventricular sleeve, said, "these cases demonstrated the life-saving potential of the device, inasmuch as it worked well at maintaining normal circulatory function during the time it was in use."

Although the device is still highly experimental, its future applications have already been spelled out by Dr. Blakemore. "Hopefully," he declares, "it may one day be called upon to take over one hundred per cent ventricular function, for as long as two weeks, to give the heart a chance to heal—after a heart attack, say, or heart surgery." Time and trauma studies in animals—and the trials in man—show this to be feasible. Also the device might be used to maintain the circulation after death, where surgeons hope to salvage healthy organs for transplantation.

Often a heart attack is not the ultimate cause of death. The body has great powers of recuperation, and the heart itself has an enormous reserve to call upon. But a complication, called shock, often follows a heart attack. Experts estimate that fully nineteen per cent of all the fatal heart attacks may be due to shock. When the heart suddenly stops beating, the arteries and veins swiftly constrict in an attempt to raise the sudden drop in blood pressure. This forces the still blood-laden veins to eject their contents into the heart with great force, overloading an already weakened pump and often making it almost impossible for it to resume its vital beating.

The traditional treatment for such cardiogenic shock has been drugs that stimulate heart action and raise the blood pressure. But six out of ten patients do not respond to drug treatment.

"The problem," says Dr. Leslie A. Kuhn of New York's Mount Sinai Hospital, "is to increase the flow of blood to the heart itself to improve its function before death intervenes. The other organs of the body, save for the brain, can survive an interrupted blood flow for a far longer period than the heart."

Using dogs, Kuhn and his associates first induced a heart attack by inserting small plastic balls into the bloodstream. Then, as the dog's blood pressure dropped sharply and it went into shock, Dr. Kuhn inserted a catheter with a balloon on the end into the femoral artery in the groin and forced it up this channel until it reached the major artery of the abdomen, the abdominal aorta. This large vessel springs from the left side of the heart and is the original spillway of all blood pumped from the heart. With the catheter in the aorta, the balloon on the end is then inflated and acts as a dam, blocking the flow of blood through the artery. As the blood backs up in the aorta, it raises the pressure and flows in all the arteries above the dam, including the coronary arteries. Rich blood flushed with oxygen, which would normally circulate throughout the entire body but which under conditions of shock is more likely to form stagnant pools in the capillaries that link the coronary arteries to the coronary veins, is instead jetted right back to the coronary arteries, force-feeding the heart with oxygen-rich blood.

Although the heart must work harder to pump blood against the increased pressure, it does so without being denied the oxygen it needs for survival.

Variations on the balloon catheter are being developed to convert it to a pump that can assist the heart through a crisis. A Detroit group has designed a balloon to massage the heart when external cardiac massage fails to restore function. Developed by Dr. Aran Johnson, a surgeon, and Bert Prisk and Donald Whitney, engineers at General Motors, the balloon is inserted into the chest cavity through a small incision at the base of the breastbone. Once the balloon is in place, it is hooked up to an air-compressor which inflates and deflates it at a controlled rate. An electrocardiograph synchronizes the rate of inflation to the desired beat of the patient's heart.

Still another balloon is under development by Dr. Adrian Kantrowitz. Under contract to the National Heart Institute, his research has produced an Intra-Aortic Pump, or IAP, which consists of a balloon, ten inches long and three quarters of an inch in diameter, with a slender gas tube laced down its center. The balloon is connected to an external control unit that like others is synchronized by an electrocardiograph machine.

The balloon is threaded through the femoral artery into the aorta. Using what is known as the counter-pulsation technique, the balloon expands as the heart relaxes and deflates as the heart contracts. Thus, after the heart has emptied its blood into the aorta and begins to relax, the balloon expands and pumps the blood through the body. At the same time, the back pressure forces blood into the coronary arteries.

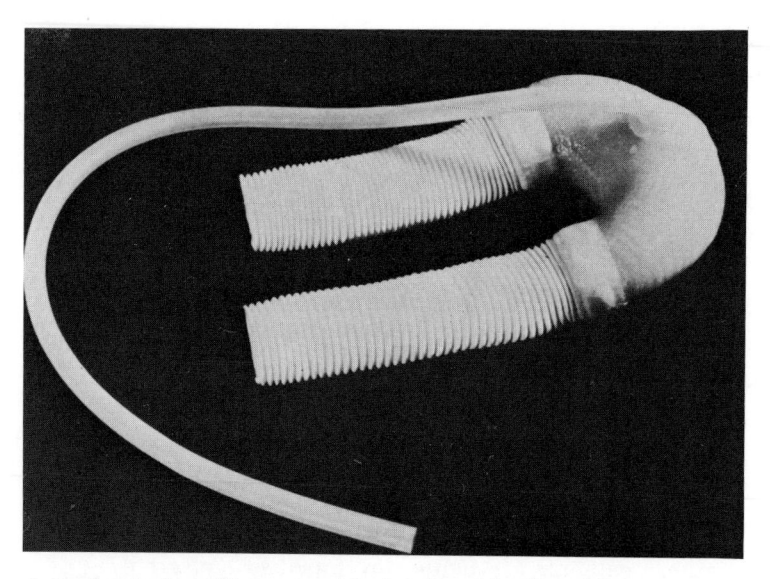

A mechanical auxiliary ventricle developed by Dr. Adrian Kantrowitz of Maimonides Medical Center and Dr. Arthur Kantrowitz of Avco Corporation. (*Avco Corporation*)

The next cycle begins as the heart starts its contraction. The balloon collapses, reducing the volume in the aorta and changing it into a low-pressure vessel into which the heart easily empties its blood.

The IAP is the most sophisticated of the balloon devices and the most experimental. It is not, however, a new area of research for Dr. Kantrowitz. With his brother, Arthur, of the Avco Everett Research Laboratory, he developed a half heart, or partial replacement, to take over the work of the left ventricle, the heart's main pumping chamber. The Kantrowitz device is a curved Dacron tube containing a flexible rubber lining and is placed entirely within the chest. It has been used clinically with some success.

The ventricle-assist idea is also being pursued by Baylor University's Dr. Michael De Bakey. Called a left-ventricular bypass pump, the De Bakey device is the size of a grapefruit and has a double lining of fabric. It is installed on the left side of the chest, with the top half protruding outside the chest wall, above the ribs. Of the first seven patients to use the De Bakey by-pass, two survived.

The future, however, seems to be not in partial-assist devices, but rather in a total replacement for the human heart. As it is obvious that the transplant of living hearts will never adequately supply the need, the answer, according to most experts, lies in the development of a working, implantable artificial heart.

"The question," says Dr. Willem Kolff, one of the pioneer inventors of artificial organs, "of whether or not such a pump inside the chest would be acceptable, desirable, and economically justifiable must be considered in the light of only one alternative—death."

Since 1957 Kolff has been trying to develop an artificial heart for man, first at the Cleveland Clinic Foundation and now at the University of Utah. The Kolff heart, which has kept calves and sheep alive for as

Dr. Adrian Kantrowitz implanting in a patient a mechanical booster pump, which takes over for the left ventricle of the heart. (*Avco Corporation*)

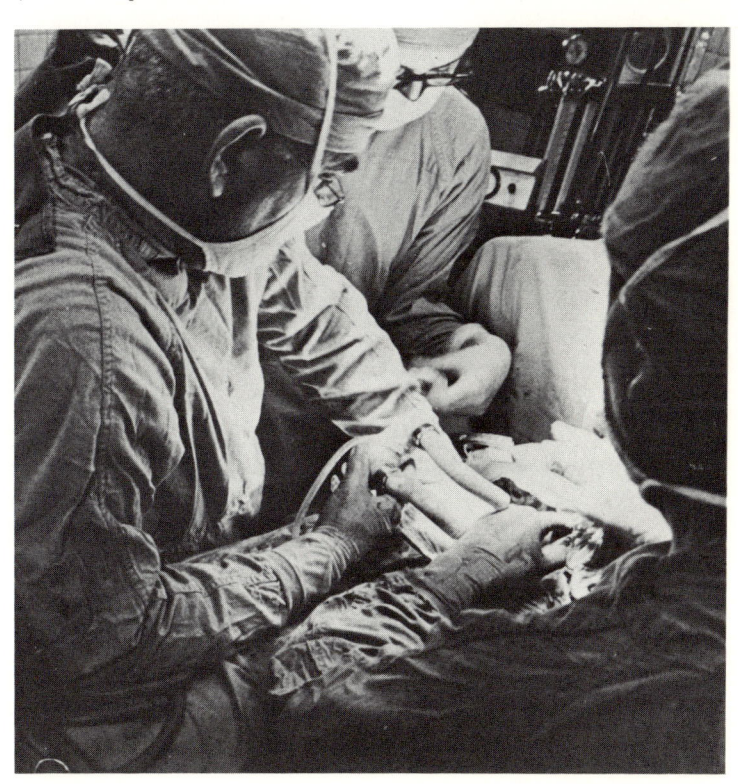

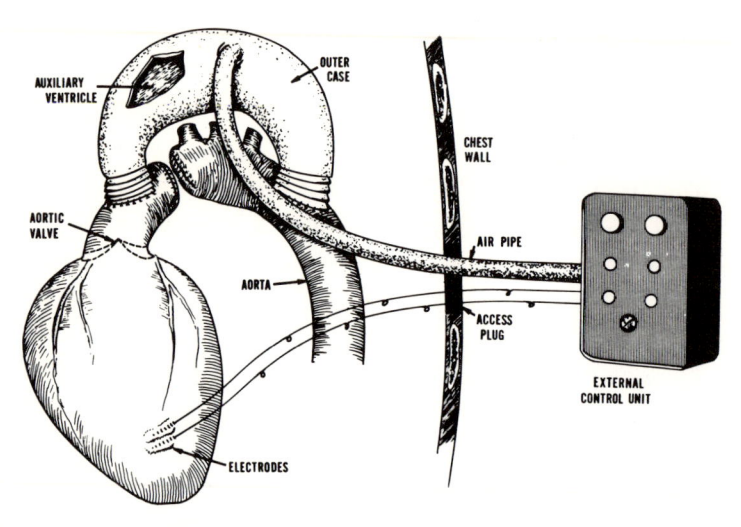

Sketch shows surgical connections that link the auxiliary ventricle to the heart. (*Avco Corporation*)

long as a day and a half, still has two major problems. "Before I would ever allow an artificial heart to be put inside a man's chest, even when his life expectancy would be short, I would have to convince myself that it wouldn't break down, so that for a reasonable time it would function satisfactorily. This is a matter of quality control, in industry . . . and this problem has to be overcome. The second problem is the problem of clotting. I personally believe that we are on the verge of really solving this problem, at least to the extent that we can live with it. It's well known that some patients that have artificial heart valves will develop clots—and some will even die from it. But far more are being saved by artificial heart valves, so you don't have to solve the problem of clot in absolute terms."

Clotting is in fact one of the major difficulties to be overcome in the development of an effective artificial heart.

Any artificial material placed into the bloodstream is soon covered with a lining of clots. Eventually tissue forms over the material and through it if it is porous, incorporating the clot into its structure. But too often, before this happens, pieces of the clot may break off and be whirled by the bloodstream back into the heart or up to the brain, causing a heart attack or a stroke.

Artificial heart valves, which have been in use for a number of years, have a known and anticipated failure rate of thirty per cent that is directly due to clotting. Long streamers of clots form about the valve at specific stagnation points and are easily broken loose and swirled about through the bloodstream. In one third of these cases, the results are fatal.

Although artificial heart valves are designed with a critical eye toward physiological function, designers apparently failed to take into account the basic principles of hydrodynamics. This appears to be the

Fabrication of plastic and rubber covering for a totally implantable mechanical heart being developed by Dr. Willem Kolff of the University of Utah. In April 1969 Dr. Denton Cooley of St. Luke's Hospital in Houston, Texas, inserted the first complete mechanical heart into a human patient. The heart functioned for sixty-five hours before being removed when a human heart was made available for transplantation.

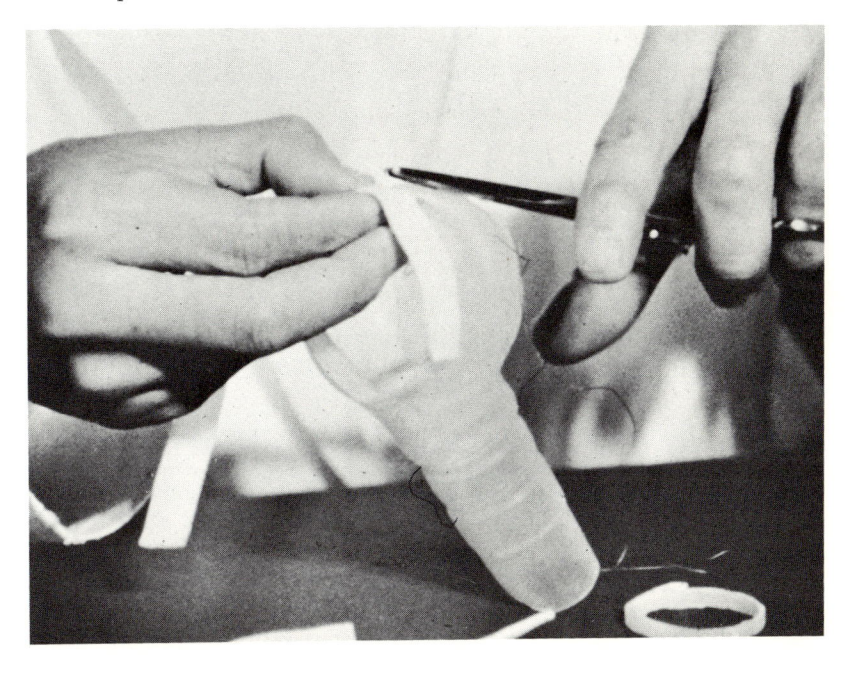

root of the problem, for clotting occurs only when the velocity of blood flow drops below certain critical speeds. Arthur Kantrowitz points out that clots show up on artificial heart valves most often in regions of stagnation or even where the flow has been reversed. He suggests that fluid mechanics, an old and well-established science, be used to engineer the proper hydrodynamic properties into the artificial heart.

To study these principles as they apply specifically to blood flow, Kantrowitz and his associates at Avco designed an experiment that might uncover some of the answers to clotting. By diverting a dog's bloodstream over a flat plate the researchers can observe the distribution of clots over the surface. The hydrodynamics of the flow was precisely calculated as the clots formed, and the Avco engineers found that clotting is the direct result of hydrodynamic changes in the blood-flow pattern.

One promising approach to the problem of clot formation on the smooth surfaces of the plastics used in artificial hearts has been explored by Dr. John Ghidoni of Baylor University. Ghidoni takes a bit of muscle tissue from a dog and grinds it up until the living cells are separated out. The cells, dissolved in solution, are then poured into a plastic aorta and with a piston-like tool rammed into the microscopic pores of the plastic. Here the cells are anchored to form a living layer over which the blood may flow without clotting.

The successful development of an artificial heart may go a long way toward solving many of the problems and issues now raised by Dr. Barnard's bold step into the arena of human heart transplantation. In a sense, this was the ultimate surgical commitment—to pluck from a human being a living heart, no matter how ill or damaged it is, seems to violate the precepts of medical ethics. And yet, it does not. The transplantation of the human heart, or its substitution by a machine, is an effort of heroic proportions, a deed that will become a commonplace event in the 21st century.

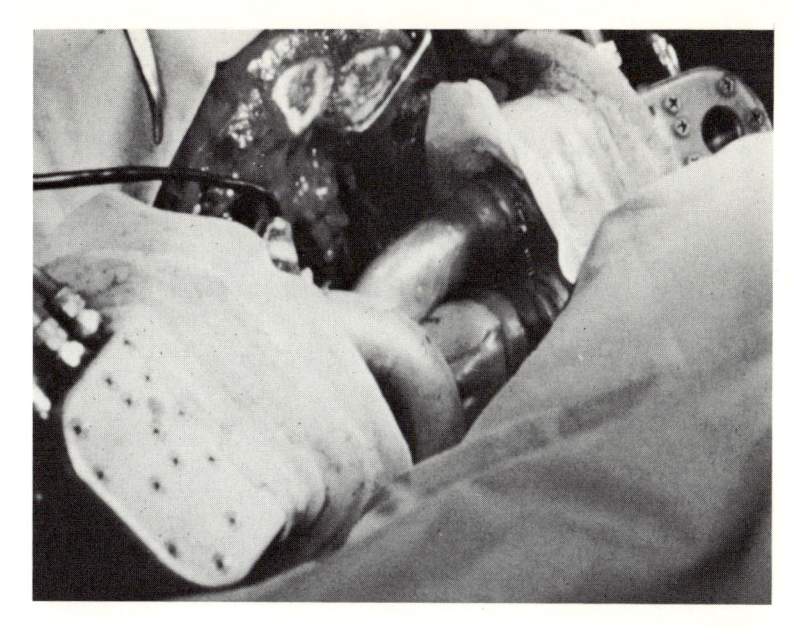

The Kolff artificial heart being inserted into the chest of a calf at the Cleveland Clinic Foundation.

The calf, twenty-four hours after receiving the artificial heart.

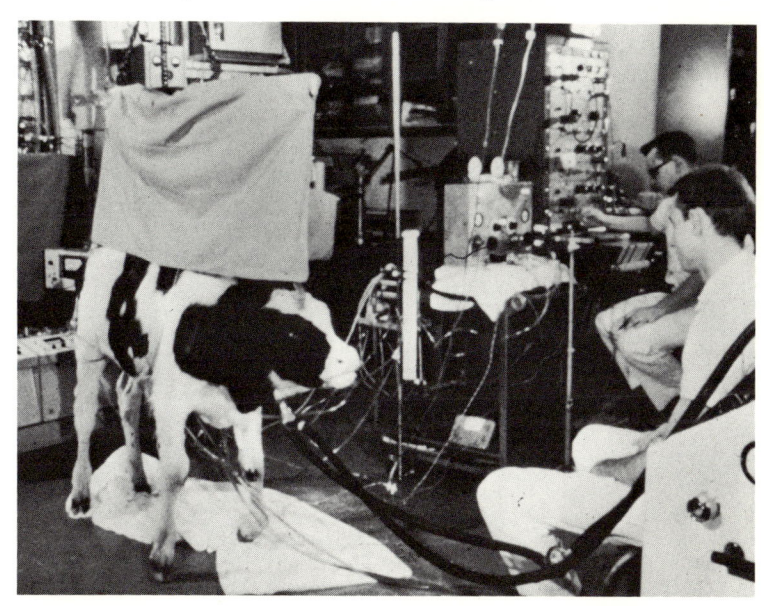

7 | Miracle of the Mind

"An enchanted loom where millions of flashing shuttles weave a dissolving pattern, always a meaningful pattern though never an abiding one. . . ." Thus did physiologist Francis Sherrington once describe the ten billion cells that constitute the human brain. Such a poetic description no longer satisfies the scientists now studying the human mind. For they have one eye on its processes and another on its control, which they consider certain to be achieved by the 21st century.

"The techniques of control," says Dr. James McConnell, a professor of psychology at the University of Michigan, "are so powerful now that I think it's true that at the moment we could take any given human being with normal intelligence and change his behavior from whatever it is now into whatever you want it to be. . . . And these techniques work. I'm convinced that they work."

The techniques of which McConnell speaks are far more sophisticated and far more dangerous than the brutish mental bludgeoning used by the Communists in the Korean War. Brainwashing, as it is called, combines physical and mental abuse to render the mind pliable and receptive to ideas that would ordinarily be repugnant. But the tools of control that McConnell and other scientists—experimental psychologists, molecular biologists, neurochemists—are developing are chemical and electronic, infinitely swifter and far more reliable.

Dr. James McConnell of the University of Michigan, whose original work with flatworms indicated that memory might be transferred chemically. "The techniques of mind control," Dr. McConnell now says, "are so powerful that we can change human behavior."

"The brave new science of the mind has already made some major advances and is on the verge of ever more significant achievements," says Dr. David Krech of the University of California at Berkeley. "I need not spell out for you what such understanding of the mind may mean in terms of control of the mind. Let us not find ourselves in the position of the atomic physicists—of being caught foolishly surprised, naïvely perplexed, and touchingly full of publicly displayed guilt at what they had wrought."

What is it the mind controllers are doing? How deeply are they probing the human mind? How great is their understanding of how men think and feel? The fact is, they have made enormous strides and produced in animals and men experimental results that are as astonishing as they are controversial.

The target of all this research is enormously complex; in fact the full extent of its complexity is only now being appreciated as it is becoming better understood.

The ancient idea of the brain as a sort of compartmentalized series of bins in which were stirred certain life forces was not so much wrong as it was simplistic. We have known for most of this century that the brain, like all matter, is made of atoms and molecules. Now, with the advent of new tools and knowledge, the brain may be explored on the molecular level and many of its complexities understood.

In its grossest form, the brain is a fairly simple structure built of three major parts. At the base is the brain stem, the knobbed outcropping that erupts from the top of the spinal cord. Nerve impulses are channeled through the spinal cord for direct action or transmission on to the brain itself. Touch a hot surface and the information is fired along a nerve to the spinal cord and converted to an action message ordering an immediate withdrawal. Your finger jerks back, without orders from the brain, even before you are aware that the finger has been burned.

But this is merely a reflex action; more complex behavior, such as breathing, heartbeat, and the generation of such primitive feelings as hunger and anger, comes from the brain stem. This brain stem, or medulla oblongata, is the old brain, and man has it in common with all other vertebrates in much the same form as it existed in the skull of the dinosaur. The brain stem handles the mechanical chores of living without instruction from the higher brain centers. As long as the brain is conscious, the medulla will maintain the delicate interplay of muscles allowing us to stand erect, for example, without any conscious effort on our part.

The brain stem may also be the governing body of our consciousness. At the University of California at Los Angeles Dr. H. W. Magoun found an arrangement of interconnected nerve cells within the brain stem that produced a particular pattern of electrochemical signals. These signals, from what Dr. Magoun called the reticular activating system, were sent to the higher brain centers that surmount the brain stem, and served as a sort of blackout switch. Only when they were sent and received was the brain aware of what was going on. When the impulses were not received by the higher brain centers, the individual lost consciousness.

Studies made during major surgery performed under a general anesthesia have shown that the nerve impulses caused by the scalpel cutting through tissue arrive at the brain stem with the same speed and intensity as they would if the patient were conscious. He does not feel them because the chemical anesthetic has turned off the consciousness switch. The pulses from pain receptors are trapped in the brain stem and are not registered in the higher centers of the brain. There is no realization of pain because consciousness has been turned off.

The recipients of these consciousness impulses sit atop the brain stem like two swollen, wrinkled pumpkins on a stick. The larger one is the cerebrum, the Latin word for brain. The smaller, tucked under and to the rear of the cerebrum, is the cerebellum, or little brain. The cerebellum is a sort of advanced model of the brain stem. What the brain stem does for the stationary body, the cerebellum does for the body in motion. Walking, running, reaching; any of the coordinated, semiwillful muscular acts are within the domain of the cerebellum. It accomplishes its control function as a result of feedback. It accepts sensory information from a variety of inputs and translates the information into specific impulses that allow us to reach for an object, say a glass, slow down the motion as the hand comes close to it, and finally stop and grasp the glass. All the automatic aspect of the maneuver is accomplished by the cerebellum acting upon the feedback information it receives from the sensory inputs fed to it.

But it is the cerebrum with its cerebral cortex, the tightly folded, surface-wrinkled outer layer of the brain, that is more highly developed in man than in any other animal. Called the new brain, it is divided into two equal hemispheres and is the site of mental activity, voluntary action, and sensory perceptions. Scientists have located specific sites along the cortex where specific functions are carried out. Sight, hearing, and the other sensory inputs are received and analyzed in fairly well-defined areas of the cortex. All the higher forms of mental activity originate here. Speech, for example, seems to be controlled on the left side of the cortex, in most right-handed people. When the speech centers are damaged, their functions can be transferred to the right hemisphere. A thirteen-year-old boy had the entire left half of his cortex removed to excise a tumor and learned to speak again.

The two cortical hemispheres of the brain are connected by an isthmus of nerve tissue called the corpus callosum. It seems to be the communications link between the two halves of the cerebrum—literally letting the right half know what the left half is doing and vice versa. When it is severed, as was done in a cat's brain in 1955 by Dr. Ronald Myers and Dr. R. W. Sperry at the University of Chicago, it was found that each hemisphere operated quite independently, as if it were a brain complete unto itself.

After cutting the corpus callosum and the optic chiasma, the crossover that sends visual information from the eyes to both hemispheres, the experimenters trained the cat to open a door while wearing a patch over one eye. In this manner, the information and the conditioning were being received by only one hemisphere. When the patch was shifted, the cat was no longer able to open the door. Moreover, it demonstrated absolutely no memory of its former training and had to be completely retrained before it could perform the same tasks with its "other" brain.

With this information, corroborated by hundreds of similar experiments, surgeons decided to cut the corpus callosum in humans afflicted with uncontrollable epilepsy. The hope was to confine a seizure to one hemisphere and thus sharply reduce its severity. The experiment with the cat convinced them that the mental faculties of the patients would not be impaired.

The surgery was first attempted in 1961 and was a complete success in terms of reducing the epileptic symptoms. Dr. Sperry, now at the California Institute of Technology, and Dr. Michael Gazzaniga have examined four of the patients postoperatively for other effects. "From the beginning," wrote Dr. Gazzaniga, "one of the most striking observations was that the operation produced no noticeable change in the patient's temperament, personality, or general intelligence. In the first case, the patient could not speak for thirty days after the operation, but he then recovered his speech. More typical was the third case: On awaking from the surgery, the patient quipped that he had a 'splitting headache,' and in his still drowsy state, he was able to repeat the tongue twister 'Peter Piper picked a peck of pickled peppers.' "

Continued study of the patients, however, revealed some definite changes in their daily behavior. If a patient brushed against a wall with the right side of his body, which is controlled by the dominant left half of the brain, he reacted spontaneously and immediately. For a long time after the operation, however, the left side of the body, controlled by the right hemisphere, did not exhibit spontaneous activity to stimulation.

The researchers then devised a series of tests to determine precisely what was taking place within the two separated hemispheres of the brain. Lights were flashed across a board in a row that spanned both the left and right visual fields. The patients reported seeing only the lights flashed in the right-hand field. When lights were flashed only in the left-hand field, the patients denied seeing any lights.

Anyone jumping to conclusions would say the patients had been rendered blind in the right hemispheres, since the left side of the visual field is normally projected to the right hemisphere of the brain and the right visual field to the left.

But when the patients were asked to point to the lights that had flashed on, they touched those that had appeared in both right and left visual fields.

"The patients' failure to report the right hemisphere's perception verbally was due to the fact that the speech centers of the brain are located in the left hemisphere," reported Dr. Gazzaniga.

The ultimate aim of the study, however, was to see exactly how the separation of the two hemispheres affected the mental capabilities of the human brain. After numerous experiments and studies, Sperry and Gazzaniga reported: "Taken together, our studies seem to demonstrate that in a split-brain situation we are really dealing with two brains, each separately capable of mental functions of a high order. This implies that the two brains should have twice as large a span of attention—that is, should be able to handle twice as much information as a normal whole brain. We have not yet tested this precisely in human patients, but . . . have found that a split-brain monkey can indeed deal with nearly twice as much information as a normal animal. We have so far determined also that brain-bisected patients can carry out two tasks as fast as a normal person can do one."

Will a combination of split-brain surgery and specialized training be used to make human beings smarter in the 21st century? A great deal more about the brain and its specific functions must be learned before that is even considered. Certainly other methods of control seem more feasible. Drugs and electrical stimulation, for example, utilize the brain's own dynamics of control. The entry of chemical agents into the field of brain research, however, is fairly recent. Virtually all brain function was thought to be electrical in nature. This idea has strong roots that reach back to the latter part of the 18th century, when Luigi Galvani first used electrical currents to send severed frog's legs into a snapping dance. Almost one hundred years later, an Englishman, Professor Richard Caton of Liverpool, pasted some electrodes onto the skulls of rabbits and monkeys and discovered that the brains gave off an electrical current.

Finally, in 1924 two great events set the trend for modern brain studies. In Austria, Hans Berger invented the electroencephalograph, the EEG, which records the brain waves on the surface of a man's skull. That same year a Swiss citizen, Walter Hess, developed a technique of implanting electrodes deep within the brains of animals in order to study individual regions of the brain.

With these two tools scientists spent the next forty years gathering information about the electrical activity of the brain and its meaning. They determined that the brain normally gives off a steady, low-voltage background pulsation called the alpha wave. This rhythm changes sharply with the mental state; alertness, lethargy, concentration, sleepiness, and specific sensory inputs will alter the pattern of the alpha wave.

The study of differing brain waves and the implantation of electrodes to ferret them out have led to a great deal of information about the brain. Control centers responsible for motor, sensory, mental, and emotional responses have all been located.

One of the early mind-mapping pioneers was Dr. Wilder Penfield of the Montreal Neurological Institute. He developed an electroprobe technique for the treatment of epilepsy. Penfield removed a section of the skull covering what he believed to be the defective brain tissue. The patient was then asked to describe his sensations as the probes were stuck into various parts of the exposed tissue and given small jolts of electricity. The surgery may sound barbaric, but it is actually quite painless, for the brain itself has no pain receptors. Only a local anesthetic was used for the entry into the skull. The idea was to determine electrically the borders of the defective tissue.

One day in 1936 Penfield was operating on a fourteen-year-old girl. The affected tissue was on the left side of the cortex, just above the ear. The probe was inserted and the current turned on. Suddenly, the girl started to describe an incident that had occurred years before. She spoke of walking through a field of tall grass, of a man carrying a sack coming up from behind. "How would you like to get into this bag with the snakes?" he asked.

All of the fear and shock of that moment were suddenly recalled and relived on the operating table with remarkable lucidity and detail. It was as if Penfield had punched the start button on a movie projector and the event, stored on some sort of memory film, was immediately projected, exactly as it had occurred years before.

It wasn't just a remembrance but was, said Penfield, "a hearing-again and seeing-again—a living-through moments of past time."

Even more startling was the fact that the patient never lost the awareness of her own surroundings. She was still on an operating table and knew it. But some hidden well of memory had been tapped and had gushed upward at the single touch of an electrical probe. Was this area of the brain, the left cortical hemisphere hard by the ear, the so-called temporal lobe, the site of memory storage? Penfield's serendipitous probing had raised a storm of questions; yet most remained unanswered or even recognized. Penfield's work was used to chart a map of the mind, a major achievement, and electrical stimulation of the brain, or ESB, became a major research tool.

In laboratories across the world the brains of living animals were being jolted with electrical current to provoke specific body movements —raise rear leg, move forward. And the animals dutifully performed as they always had, running through mazes, pressing levers, being rewarded and punished—but now accompanied by electrodes leading to an encephalograph. In the process, the brain was being mapped as never before. Physical movement could be assigned to a specific part of the brain. Then, with a newly minted Ph.D. in psychology, James Olds, working at McGill University in Montreal in 1953, was taught the technique of implanting miniature electrodes in the brain of the rat.

"So I went up to the lab one Sunday afternoon," he recalls, "and took the first rat I had ever prepared with my own hands. Every time the rat walked into one corner of the testing table, I turned on the electricity to see if he would avoid approaching that spot thereafter. Instead, my rat *liked* it!"

Just as Penfield had found a memory-storage area in the mind, Dr. Olds had uncovered a site within the brain that responded with intense pleasure to electrical stimulation. This experiment has been carried out with variations in hundreds of laboratories across the world. There is no doubt that a pleasure center exists within the brain and that laboratory animals would rather receive a few minivolts of electricity directed to this area than eat, sleep, make love, or engage in any other animal activity. In one experiment, a rat was taught to press a self-stimulation lever. Each time he pressed it, a brief burst of electricity was triggered in his pleasure center. Set about him in the cage were all sorts of rat goodies, designed to distract and engage him. The rat would have none of them. Instead, he pushed the self-stimulation lever five thousand times within the space of one hour.

Animals wired for pleasure seem to do as well on standard tests when the stimulation is offered as a reward for learning as do animals whose rewards are merely food. Might such techniques ever be applied to people? Would they work? is perhaps a better question at this point. Years of experimentation in many fields of science have demonstrated the errors that may result from directly extrapolating results obtained with animals to man. Still, in the higher species studied, ESB seems to work without exception.

One of the most dramatic demonstrations of its power was given by Dr. José Delgado, a medical doctor and neurophysiologist at the Yale University School of Medicine. Delgado has made ESB wireless—he sends electrical impulses by radio to the electrodes implanted in the brains of his experimental animals.

In 1964 he entered a bull ring in Madrid with a small radio transmitter in his hands instead of the more traditional *muleta*. A fighting bull came charging out of the *toril* at Dr. Delgado. Unlike other bulls, however, this one had a tiny electrode planted in what Dr. Delgado hoped was the aggression center of its brain. Delgado sent a radio pulse into

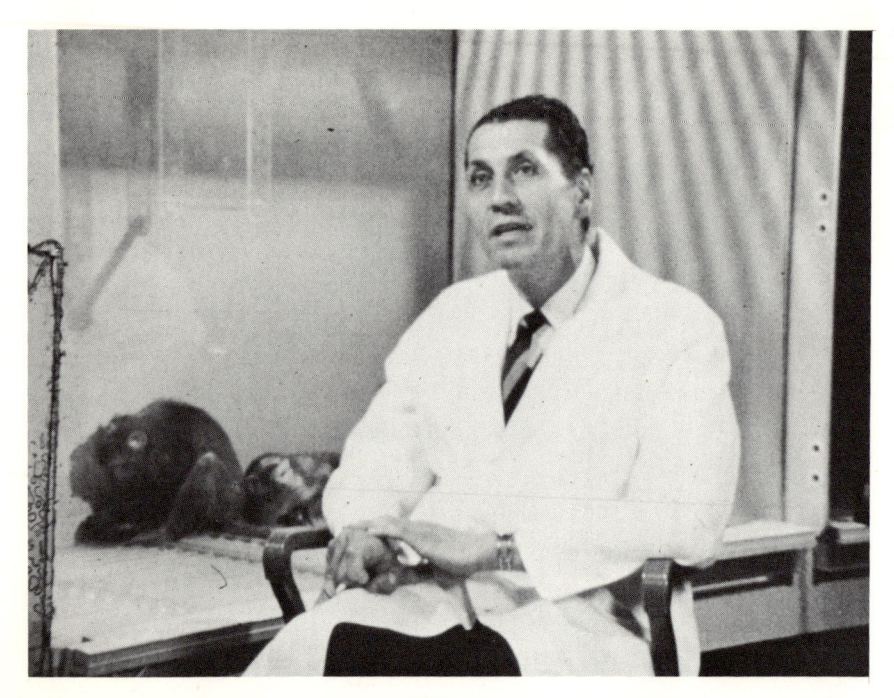

At Yale University, Dr. José Delgado produces behavioral changes in monkeys by sending radio signals directly into their brains.

the electrode. The bull stopped in mid-charge—reduced from a *toro bravo* to a tame Ferdinand—and trotted docilely off in a new direction.

"What can we do when we try to stimulate the brain?" asks Dr. Delgado rhetorically. "There is a wide range of reactions that we can initiate: We can modify the heart rate, the respiration, gastrointestinal activity, we can induce movements, flexion or extension of the arms, movement of the head—all kinds of motor activity."

By careful placement of the electrodes, Delgado can stimulate an animal to turn circles, endlessly moving around and around, ignoring everything else about him but the need to keep up the strange but compelling circular movement. Dr. Delgado and other experimenters have found the site of hunger and satiation in the monkey brain. By stimulating one area, the experimenter can cause monkeys, already full to bursting, to keep stuffing bananas or food pellets into their mouths until they are so incredibly filled that they must regurgitate what they have

The technique called electrical stimulation of the brain has been used not only to modify individual behavior but also to change the entire social order of a monkey colony.

just eaten in order to keep forcing the food down their throats. But as long as the impulses jab at the brain, they will continue to eat. Similarly, an electrode placed in the satiation center will keep an animal from eating. Food has been placed before animals on the brink of starvation, but nothing can induce them to eat so long as the appropriate center of the brain is being stimulated.

By eliminating the wires that connected the animal to the source of the stimulation, Dr. Delgado has turned them loose without losing the ability to send tiny shocks of electricity into preselected areas of the brain.

"If we use radio instrumentation," he explains, "the animals are completely free, and we can investigate not only the individual but also social behavior."

With rhesus monkeys, Dr. Delgado not only studies but also modifies individual and social behavior. In the laboratory one monkey, by virtue

155

Careful mapping of the various centers of the brain enables researchers to implant the electrodes in those areas that produce specific behavior. The hunger center can be turned off or on so that sated monkeys will continue to jam food into their mouths and famished ones will refuse all food. Aggression can be turned off and on, as can pleasure, by electrical stimulation of the brain.

of his size and personality—usually mean and aggressive—is dominant. He takes over a desirable part of the cage and sits glowering at all who approach. The other monkeys in the cage, each aggressive and dangerous in its own behavior, recognize the unseen boundaries the Boss Monkey has staked out. If one of the lesser monkeys violates the Boss's territory, the Boss bares his teeth, snarls, and barks his displeasure. The intruder beats a hasty retreat.

This is the classic pattern of behavior in the cage, with the Boss Monkey, the dominant male, taking his pick of the food, females, and territory. A radio signal to the electrode implanted in the Boss Monkey's skull can drastically alter that pattern. The impulse abruptly changes the Boss's personality. He becomes immediately docile; no longer does he scowl at the other, lesser, occupants of his cage. His behavior has been altered by electrical stimulation, and as a result the entire social structure of the small tribal group is also changed. But after the signal stops, the original situation is restored. The Boss Monkey resumes his scowling, dominant ways, shattering the new societal structure and restoring the original one.

For Dr. Delgado, the most interesting aspect of this work is the possibility of influencing the physiological functions of the brain. "By this, I mean that we can induce pain and pleasure. We can modify aggressive behavior. . . . We can make our animals more aggressive or less aggressive. We can influence nocturnal behavior. Through electrical stimulation of the brain in humans, we can modify even the thinking process."

Might a 21st-century version of such techniques be used to control the behavior of a human population in the future? Delgado does not think so. "Fortunately," he says, "the prospect is remote, if not impossible, not only for obvious ethical reasons, but also because of its impracticability. . . . This technique requires specialized knowledge, refined skills, and a detailed and complex exploration in each individual. . . . The application of intracerebral electrodes in man will probably remain highly individualized and restricted to medical practice."

One clinical approach has been made at the Tulane University Medical School. Dr. Robert Heath has buried electrodes in the brains of violent psychotics. Wires lead from the electrodes to battery packs and control buttons, which are carried by the patients. Each time they feel rage, depression, or some other violent emotion overtaking them, they simply press the button and are gently shocked into a pleasurable lolling state.

The technique is very primitive and not really promising at this point. Electrodes cannot be left in the brain for very long periods of time. The response to stimulation is also brief, and the result is simply a temporary easing of a bad condition, not a viable therapy.

But these methods are not meant to be anything but research, a means of determining the techniques needed to help specific individuals with specific illnesses that seemingly cannot be treated in any other way. Control of the mind implies far more than this. The idea of even the slightest possibility of altering human behavior as a means of solving some of the seemingly insoluble human problems is too alluring to be dismissed lightly. However, scarcely anyone in the field of brain research will talk of control as the ultimate goal; most researchers speak of understanding the human mind and so solving the problems it creates. Chief among them is war. Why do men fight—individually and as nations? We have no answers, of course, but there are indications that control might be possible on the basis of some experiments already reported in the literature. These concern the aggression center of the brain. Dr. K. E. Moyer of the University of Pittsburgh described one such experiment, performed not on an animal but on a human being, at a recent UNESCO meeting in Paris. He told of a shy, even-tempered woman patient under treatment for an organic disorder. Electroprobes were placed in her brain and nothing untoward occurred. But when the current was sent into a cluster of nerve cells called the amygdala in the temporal lobe, the woman became violent. She cursed the surgeon and threatened him with thrashing arms. When the stimulation was turned off, she tumbled back immediately into her former gentility. She recalled exactly what had happened and apologized.

"Her hostile feelings and aggressive behavior could be turned on and off at the flick of a switch," explained Dr. Moyer.

Who will operate those switches in the future and for what purpose? Most of the mind researchers dismiss the idea that some future dictator will ever utilize mind-control techniques such as ESB, but there is no gainsaying that mind control is a reality now and its practice limited only because it is still so new. There is also no reason to suppose that a technology capable of putting men on the moon or transplanting human hearts will not one day solve the technical problems associated with electrical stimulation of the human brain.

Electrodes may not even have to be placed in the brain. At the Space Biology Laboratory of the University of California at Los Angeles Brain Research Institute, Dr. W. Ross Adey has been examining the effect of external force fields upon the human brain. He is convinced that it can respond to electrical stimuli far below the levels that are normally thought to be effectual. Analyses of the brain waves given off by chimpanzees when they perform some learned task have been made, and Adey can distinguish in brain-wave patterns those that relate to

At the University of California at Los Angeles Brain Research Institute, Dr. W. Ross Adey examines the effects of external force fields directed to the brain without implanting electrodes. Dr. Adey believes that ambient electrical energy might increase learning by altering mood.

correct decisions and those that accompany incorrect decisions. With this as a basis, he thinks that it might be possible to control the rate at which the mind learns. "I think, though we have no means of desirably affecting the learning rate at this time, there is every reason to think that we will be able to do something about it.... The intention that one might have in applying force fields outside the head to alter behavior

159

would be to achieve sets of mood or sets of attention under which the individual would remember faster and better the data that was being fed in," he says.

One of the experiments in this area is to place a subject between two large metal plates that generate about twenty cycles of energy. Unfortunately, the effect does not enhance learning, but rather upsets the ability to concentrate. In fifty people subjected to the experiment, all performed less well on such tasks as estimating elapsed time and recognizing a tone than they did when not subjected to the energy field.

But this does not rule out the use of external electrical stimulation to control the mind for one purpose or another. The proper energy levels and the manner in which they are radiated must be developed. In time they probably will be.

"My personal concern," says Dr. Adey, "is that we do it well, that if

Dr. Jan Berkhout of the University of California at Los Angeles is attempting to measure the effects of emotional stress on the brain. He tapes a series of questions designed to cause emotional stress upon a volunteer subject.

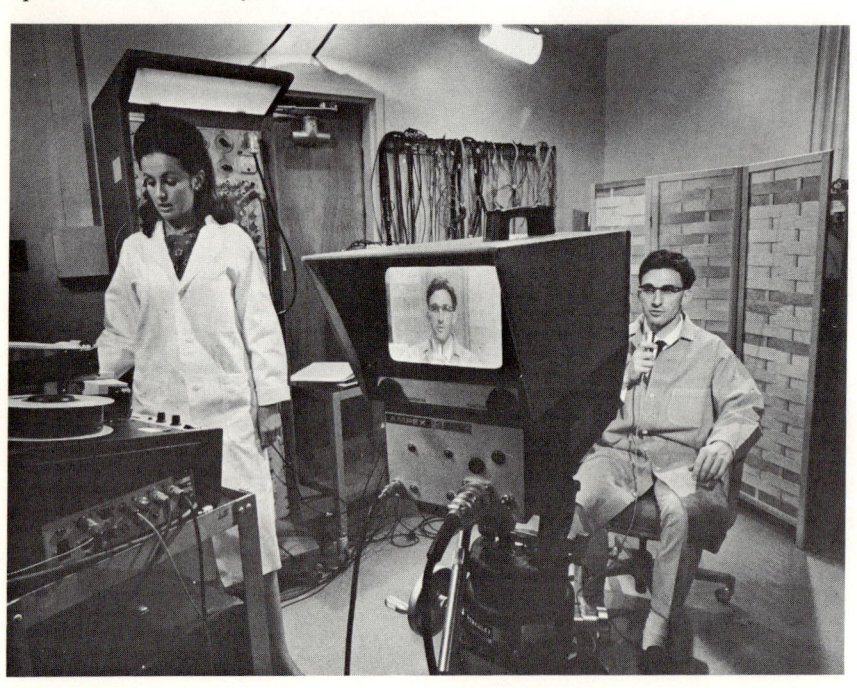

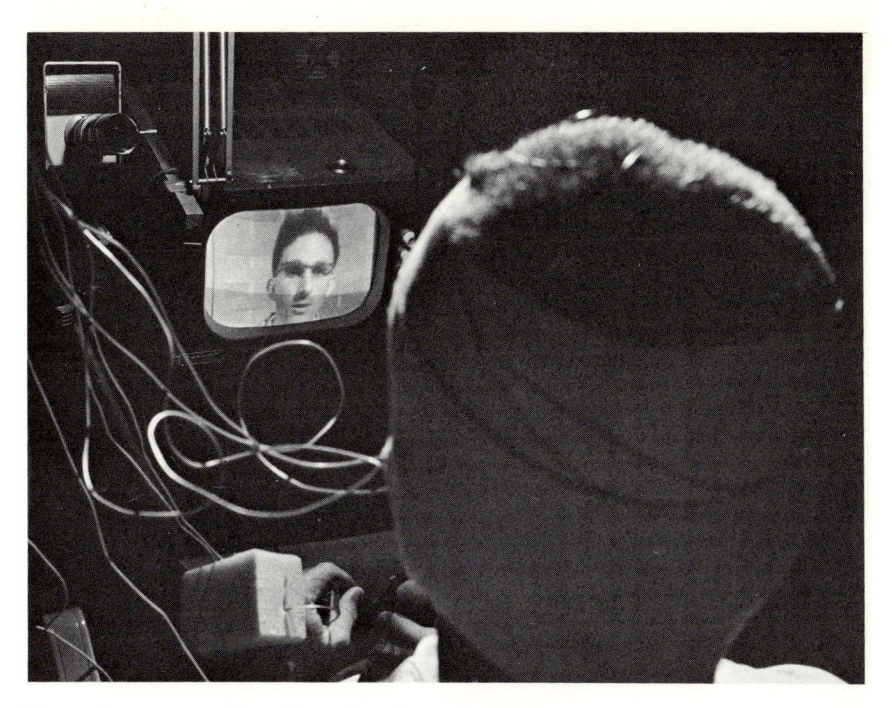

The volunteer, enclosed in a booth and wired to an electroencephalograph, answers the questions while a TV monitor transmits his answers to Dr. Berkhout.

we decide that this manipulation is feasible, that we do it in ways that are socially acceptable. I think the current public interest in drug manipulations in this regard is rather overoptimistic. My feeling is that we will learn the patterns of the environment that optimize learning rather than using drugs for internal manipulation."

Drugs and chemical agents have been playing an increasingly large role in brain research in recent years. For despite the startling effects achieved by electrical stimulation of the brain, there had evolved a growing feeling that electricity alone could not explain the complex performance of which the brain was capable. Memory and learning, for example, are common to all animals that possess brains, but no wholly satisfactory explanation of these processes had been made before 1960. Even now there are arguments about specifics, but a good deal has been learned and a number of long-extant theories have been overthrown.

While the tape continues to roll and ask the questions, thus assuring
the same phrasing and intonation for each subject, Dr. Berkhout
examines the EEG. From these studies the effects of emotional stress
on brain chemistry and behavior may be determined.

The processes of learning and the storage of experience, that is to say memory, were long thought to be electrical. Repetition of a given act or of a particular stimulus was believed to produce a specific pattern of electrical activity in a given circuit of neurons. The pattern was, according to the theory, self-perpetuating, reinforced and brought back with increasing ease each time the original stimulus was repeated.

The electrical theory received a sharp jolt, however, when brain researchers began to take carefully trained animals, in whom specific responses could be elicited by the presentation of specific stimuli, and subjecting them to a battering of what should have been electricity-destroying shocks. Powerful drugs, glucose starvation after insulin injections, electroconvulsive shock treatments, and other agents designed to disrupt temporarily all brain-wave patterns were used. But in each case the animals, after recovering from the ordeal, remembered their training. One experiment even tried to freeze the memory out of carefully trained mice by placing them in a chamber and then dropping the temperature below freezing. At this point virtually all brain-wave activity ceased, but when the mice were thawed out, they retained their memories. Something more permanent than neural electricity was the storage bank of memories. It was at this point that molecular biology, the new discipline that had established DNA as the chemical molecule responsible for the transmission of inherited characteristics, entered the picture.

The main problem in dealing with the chemistry of the brain had been the scientists' inability to focus on individual brain cells. The brain is composed of some ten billion neurons, or nerve cells. A single neuron consists of a main cell body, in which is stored the nucleus with its DNA-coded genes awash in the fluid of the cell and the other basic components common to most cells. Extending out from the central cell body are an axon and several dendrites. External stimuli are received by the dendrites, or input leads, converted into electrical impulses, and flashed to the cell body. If the stimulus is of sufficient strength to pass the threshold, or smallest quantity of electrons needed to activate the cell body, it fires, passing an electrical signal along the axon, or output line. Here, the signal leaps to a junction called the synapse, which connects the axon with the dendrite of another nerve.

According to a new theory, this elegant electrochemical system is governed by a number of very specific proteins and the templates on which the proteins are made, the large molecules of ribonucleic acid. The RNA molecules in turn are manufactured under the specific direction of the DNA, the deoxyribonucleic acid produced in the main nerve-cell body and in cells closely associated with it called glial cells. But it was not until 1957, when the technique of isolating one single nerve cell from the thousands of others to which it was interconnected was developed, that this electrochemical theory, not only of nerve impulse transmissions but also of memory and learning, was first elucidated.

At the University of Göteborg a Swedish biologist, Holgar Hydén attacked the problem by designing a minute set of surgical tools—stainless-steel picks, knives, hooks, and probes—all scaled down to the micro-size needed to tear a single neuron free of the brain. Next he had to train his hands to use these tools under a binocular microscope.

Finally, he was able to strip down a neuron the way a housewife peels a carrot, slicing off the glial cells that stick to the neurons like Band-aids and then plucking the nucleus from the single nerve-cell body. Thus each separate part could be analyzed for its biochemical properties. How much RNA did the cell make? What kind was it? Was it always the same? Did the nucleus provide only DNA or did it also make RNA? And what of the glial cells? Were they there as structural components or did they perform a specific biochemical function?

By 1960 Hydén and his co-workers had a number of startling answers. They trained rats and other animals to perform certain tasks. Some tasks were simple, others more complex. Immediately after performing a task, or receiving some sensory stimulation, the animals were sacrificed and the individual neurons that made up their brains were studied. The results demonstrated for the first time a tentative chemical basis for learning and memory. Stimulation of any sort, Hydén found, increased the rate of RNA production inside the neurons. This, of course, meant that protein was being manufactured at a greater rate as well, for RNA serves as template upon which protein is built. In addition, as the RNA rate of manufacture increased in the neurons, it declined in the glial cells that clung like satellites to the neurons. It appeared in fact that the glials were actually siphoning off their RNA to the neuron for use during peak periods. When the nerve stops firing, the glials again make RNA for their own use.

As the role of RNA became increasingly implicated, an astonishing fact was noted. Every training session not only caused more RNA to be produced but also caused a slight difference in the make-up of the RNA. The chemical bases of RNA in its molecular form can be changed or rearranged into trillions of different combinations, each varying slightly from the next, with the result that each RNA molecule can bring about the manufacture of a protein that differs slightly from all others.

And so it was with the animals Hydén studied. Those that had undergone training seemed to be yielding neurons that produced RNA molecules slightly different in their base structure from the RNA taken from the neurons of control animals that had received no training.

Furthermore, it was noted that the RNA changed slightly during the course of an experiment. Molecules extracted early in the training period differed slightly in their base sequences from those produced toward the end of a learning experience.

From a mountain of experimental data, Hydén put forth the theory that sensory impulses activate the neural DNA to produce RNA. This in turn leads to the formation of specific proteins that store and reproduce memory. When triggered, the memory is then transmitted from cell body to axon, across the synapse, and on to the dendrites of another cell.

Support for Hydén's theories has come from a number of diverse sources—researchers with a variety of experimental approaches to the same problem. At the University of California at Berkeley, Dr. David Krech and Dr. Mark Rosenzweig performed a now classic experiment. One group of infant rats was placed in an intellectual slum—solitary confinement in a dimly lighted room. Another group from the same litter was placed in an enriched environment—cages filled with mazes, tunnels, ladders, and other intellect-sparking toys. As the rats grew, those in the enriched environment were given tasks—mazes to run, light and dark cards to choose. The impoverished-environment group remained unstimulated.

After eighty days both groups of rats were sacrificed and the brains analyzed. The enriched-environment rats had cortexes that averaged 4.6 per cent heavier than those of their impoverished litter mates. The enriched group had also produced greater amounts of cholinesterase, indicating that the synaptic pathways had been well traveled. The enriched rats had more glial cells and longer and more numerous dendrites than the unstimulated group.

Among the most controversial of the chemical memory experiments was that originally performed by Professor James McConnell of the

University of Michigan. McConnell took very primitive creatures, the flatworms known as planarians, and taught them to do certain tricks. His educated planarians learned to contract when a light went on. The principal teaching device was an electric shock that the worms received just after the light went on. After a time the worms learned to contract each time the light went on, even without the jolt of electricity. The time it took the worms to learn the task was dutifully recorded. They were then chopped up and fed to a group of ignorant—that is, untrained—planarians.

These little cannibals were then taught the same trick of contracting their bodies when the light went on. But they learned it much more rapidly than the original group of worms and more rapidly than a control group of worms that had been fed untrained victims.

The conclusion McConnell drew from this and related experiments was that specific memory of the training was being transferred by RNA molecules contained within the chopped-up worms. The experiment and its conclusion kicked up a violent storm of controversy among brain researchers. Some experimenters tried to reproduce the experiment and failed to get any speed up in the learning rate of their cannibalistic worms. Still others tried the same form of experiment on higher animals, rats, mice, and hamsters, and claimed the same type of results as McConnell got. Some even reported the transfer effect could be obtained between different species. Train a rat to do a trick, feed its brain cells to a mouse and the mouse would learn the same trick at a faster rate than a mouse on a different diet.

The controversy continued for five years, with much experimental evidence being reported to support the chemical transfer idea and some negative results sent in by researchers who simply could not reproduce McConnell's results.

Then, in a determined effort to prove or disprove the McConnell thesis, researchers in nine laboratories in the United States and abroad set up experimental models in the hope of transferring learning and memory by chemical extracts.

The most successful experiment was one perfomed by Dr. Richard Gay and Dr. Alfred Raphaelson of Flint College in Michigan. They trained rats, whose normal preference for a hiding hole would be a darkened box, to prefer a flood-lit box. The conditioning was, as usual, the electric-shock technique, and at the same time a control group of rats was given free choice in the selection of a box.

After enough training sessions had produced preference for the bright lights upon the experimental group, while the control group naturally indulged the dark sides of their natures, they were all sacrificed.

The brains of both groups were reduced to a substance that consisted largely of RNA. These brain RNA extracts were then injected into nonconditioned rats. Those that got the extracts from the untrained members of the control group followed their natural proclivities, spending a mean time of 120 seconds in the dark box during a 180-second test. The animals receiving the brain extracts from the shock-trained rats acted as if they too had received the training. The mean time in the dark box was a scant fifteen seconds, and five of the eight rats in the experimental group flatly refused even to enter the darkened box.

This experiment was duplicated successfully by several other researchers, thus making believers out of the skeptics. Planarians, one of them pointed out, are a very difficult experimental animal to work with, and many of the failures by researchers to reproduce McConnell's results might well have been due to their inability to train the flatworms properly.

But the rats were familiar experimental animals, with well-known habits and peculiarities. The same Raphaelson-Gay experiment was reproduced by a number of others. One of them, Dr. George Ungar, professor of pharmacology at Baylor, got even better results. Ungar flatly attributed them to a chemical transfer of learning. "The only possible explanation," he notes, "is that during learning, increasing amounts of some specific substance are formed in the brain. This substance must contain in its molecular structure the information acquired during learning. The information thus coded, when introduced into the recipients, induces a change in behavior."

The question the transfer people must now resolve is the exact nature of the memory molecule. Is it RNA? There is some doubt about this, for Ungar got his results from a brain extract that had been exposed to an RNA-destroying enzyme. Thus, another theory holds that the RNA merely provides the templates for the construction of protein molecules that are the actual storage bins of memory.

Such memory molecules may not provide the entire explanation of how the brain remembers. Memory appears to be a three-tiered arrangement. When a task is first learned, it seems to be stored within the rather evanescent electrical impulse that flits along the neural lines in the brain. This short-term memory can be quickly erased by electroshock techniques. In experiments on rats, chicks, and other experimental animals, learned tasks are immediately forgotten when an electroshock is applied immediately after the task has been learned. In human beings, the same results have been noticed among people receiving electroshock treatments for psychiatric problems. Invariably, they report an inability to remember experiences they had just before the treatment.

The storage mechanism for short-term memory is apparently quite limited. A phone number looked up one moment will be forgotten almost immediately if it is not repeated several times. There is a built-in dissipation factor in short-term memory.

This memory decay led researchers to assume that there is an intermediate memory stage, taking over at some time before the thought has been forgotten and before it has been consolidated as a permanent memory. This intermediate memory can also be jarred loose by shock.

The third stage, that of permanent memory, is when the actual storage or consolidation takes place. This event is not electrical; electroshock and every other technique known to destroy or interrupt electrical currents fail to destroy long-term memory. It is virtually certain now that this form of memory is the result of some chemical change within the brain. To some researchers, this change is the manufacture of slightly different RNA.

Attempts to utilize this idea and at the same time to prove it have been made not merely on experimental animals such as planarians but also on human beings. One highly publicized trial was made by the late Dr. Ewen Cameron at the Albany, New York, Medical Center. Cameron gave a drug called magnesium pemoline to a group of senile patients. He then tested their ability to remember by having them look at a geometric drawing and then copy it a few moments later. The patients receiving the drug did noticeably better than a control group of senile oldsters.

Unfortunately, the memory pill, as it was dubbed, did not produce permanent changes in the patients' ability to remember. Nor did it noticeably increase RNA synthesis. Rather, it seemed to act as a stimulant to the central nervous system, a kind of swift memory jog that was just as swiftly lost.

These and dozens of similar experiments with drugs that enhanced RNA production or acted simply as stimulants have not resolved the controversy regarding RNA versus some other mechanism.

One group believes that RNA synthesis is merely an intermediate stage, that RNA serves as a messenger for the construction of a new type of protein in which the memory is stored or encoded.

"In our laboratory at the University of Michigan we have demonstrated that there is a connection between the consolidation of memory

At the University of Michigan, Dr. Bernard Agranoff trains goldfish to swim from one end of the tank to another when a light goes on.

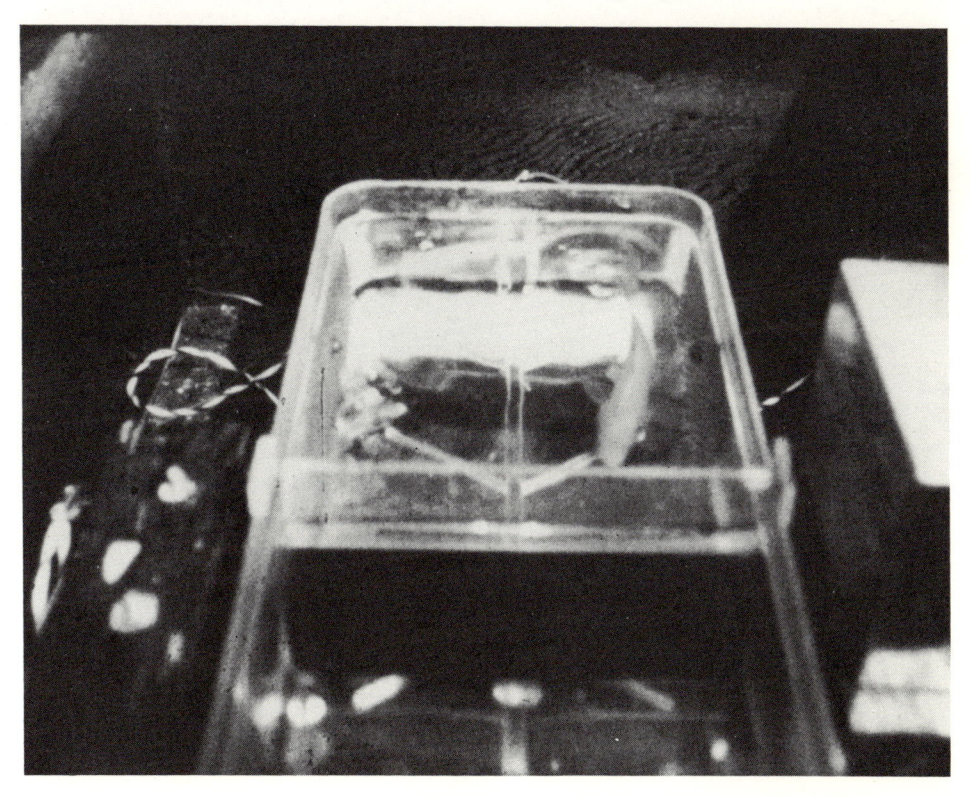

Immediately after the goldfish demonstrates that it has learned the task, Dr. Agranoff injects it with an antibiotic. The goldfish is then returned to the tank and once again challenged with the light. After the injection, there is no response, as if all memory of the training has been wiped clean.

and the manufacture of protein in the brain," declares Dr. Bernard Agranoff.

Agranoff trains goldfish to swim from one end of a tank to the other when a light goes on. He then injects an antibiotic called puromycin directly into the skull of the fish. Puromycin has a specific effect. It stops the growth of the long chain that makes up a protein. It had a very specific effect on the experiment. Dr. Agranoff explains: "We found that if the puromycin was injected immediately after training, memory of the training was obliterated. If the same amount of the drug was injected an hour after training, on the other hand, memory was unaffected."

Puromycin has no effect on RNA, only on a protein. Still, Agranoff is not prepared to admit that RNA is not the factor. The precise mechanism is still not known, and both the protein and the RNA adherents have not exhausted all experimental methods of proving their ideas. In one sense, it almost does not matter who is right.

Whatever the actual configuration of the memory molecule, there seems little doubt that learning and memory in all their subtle complexities will be much better understood in the not too distant future. With this understanding will come the possibility of controlling human behavior to a degree never before dreamed of, much less attempted. The purpose of such control and the degree to which it is exercised will have a greater effect upon the future of mankind than any other single scientific advance in the history of man. It might make the 21st century a new golden age of the mind, or it might plunge the individual into the depths of a permanent nightmare from which he will never emerge. The choice may well be ours to make right now.

SUGGESTIONS FOR FURTHER READING

1. The Creation of Life

Oparin, A. I. *Origin of Life*. New York: Dover Publications, 1953.
Sullivan, Walter. *We Are Not Alone*. New York: McGraw-Hill, 1964.
The Scientific Endeavor. New York: The Rockefeller Institute Press, 1965.
Haldane, J. B. S. *The Inequality of Man*. London: Penguin Books, 1937.

2. The Living Message

Gardner, John E. *Principles of Genetics*. New York: John Wiley and Sons, 1968.
Engel, Leonard. *The New Genetics*. Garden City, N.Y.: Doubleday, 1967.
Borek, Ernest. *The Code of Life*. New York: Columbia University Press, 1965.
Beadle, George and Muriel. *The Language of Life: An Introduction to the Science of Genetics*. Garden City, N.Y.: Doubleday, 1966.
Dobzhansky, Theodosius. *Evolution, Genetics and Man*. New York: John Wiley and Sons, 1966.
Watson, James D. *The Double Helix*. New York: Atheneum, 1968.

3. The First Ten Months

Liley, H. M. I., with Day, Beth. *Modern Motherhood*. New York. Random House, 1966.
Gilchrist, Francis G. *A Survey of Embryology*. New York: McGraw-Hill, 1968.
Berrill, N. J. *The Person in the Womb*. New York: Dodd Mead, 1968.
Balinsky, B. I. *An Introduction to Embryology*. Philadelphia: W. B. Saunders Co., 1965.
Gilbert, Margaret Shea. *Biography of the Unborn*. New York: Hafner, 1962.

Berrill, N. J. *Growth, Development and Pattern.* San Francisco: W. H. Freeman, 1961.

4. Standing Room Only

Guttmacher, Alan F. *Babies by Choice or Chance.* Garden City, N.Y.: Doubleday, 1959.

Smith, Kenneth. *The Malthusian Controversy.* London: Routledge and Kegan Paul, 1951.

Rock, John. *The Time Has Come: A Catholic Doctor's Proposals to End the Battle over Birth Control.* New York: Knopf, 1963.

Michelmore, Susan. *Sexual Reproduction.* Garden City, N.Y.: Natural History Press, 1965.

5. Man-Made Man

Warshofsky, Fred. *The Rebuilt Man.* New York: T. Y. Crowell, 1965.

Moore, Francis. *Give and Take.* Garden City, N.Y.: Doubleday, 1964.

O'Donnell, T. J. *Morals in Medicine.* Westminster, Maryland. Newman Press, 1959.

Schmeck, Harold. *The Semi-Artificial Man.* New York: Walker and Company, 1965.

6. The Human Heart

Smith, Anthony. *The Body.* New York: Walker and Company, 1968.

Blakeslee, Alton, and Stammler, Jeremiah. *Your Heart Has Nine Lives.* New York: Pocket Books, 1964.

Gertler, Menard. *You Can Predict Your Heart Attack and Prevent It.* New York: Random House, 1963.

Asimov, Isaac. *The Human Body.* Boston: Houghton Mifflin, 1963.

Lamb, Lawrence E. *Your Heart and How to Live with It.* New York: Viking, 1969.

7. Miracle of the Mind

Wooldridge, Dean E. *The Machinery of the Brain.* New York: McGraw-Hill, 1963.

Wilentz, Joan. *The Senses of Man.* New York: T. Y. Crowell, 1968.

McGaugh, James L., et al. *Psychobiology.* San Francisco: W. H. Freeman, 1967.

INDEX

Abiogenesis, doctrine of, *see* Spontaneous generation
Adamsons, Karlis, 76
Adenosine triphosphate (ATP), 17
Adey, W. Ross, 158-60
Agranoff, Bernard, 169-71
ALG (antilymphocyte globulin): use of, to prevent rejection of transplants, 111-12
Alpha-adrenergic blocking agents, 135
Amino acids, 6*ff*, 29-32; *see also* Chemical evolution of life, theories of
Amniocentisis, 66-67
Amniotic tap, 63
Animal populations: safety check to prevent overpopulation of, 82-87
Antibodies, 106-109; formation of, 106-109; RH disease and, 61, 64-66
Antibodies and antigens: clonal-selection theory of, 107; instructive theory of, 107
Antigens, 107, 111, 113; *see also* Antibodies
Antilymphocyte globulin, *see* ALG
ATP, *see* Adenosine triphosphate
Avco Everett Research Laboratory, 138-140, 142

Bailey, Charles, 130
Balloon catheters used in the treatment of heart disease, 136-38
Barghoorn, Elso, 16
Barnard, Christiaan, 123-25, 142
Barr bodies (female sex chromatins), 71-72
"Battered-child syndrome," 86
"Behavioral sink," 86
Behrman, Richard, 76
Berger, Hans, 151
Berkhout, Jan, 160, 161, 162
Bernal, J. D., 9
Berrill, N. J., 50
Bevis, Douglas, 63
"Biological marker," 16
Birth control: attitude toward, in India, 90; encyclical of the Pope on, 90-91; in the future, 99-100; in Korea, 80-81; *see also* Contraceptives
Birth-control capsule, long-acting, 97-99
Birth-control pill: action of, 94-95; drawbacks of, 95
Birth defects and infant mortality, 45-46
Blaiberg, Philip: longest survivor of heart transplant, 123, 125
Blakemore, William, 136
Bloch, J. H., 120
Blood: deep-freeze of, 119
Blood banks, 118

175

Blood clots: development of drug that will form soft clots, 133; hydrodynamic changes in blood-flow patterns as cause of, 141-42; as the major problem in the use of artificial hearts and heart valves, 140-42; role of, in heart disease, 131-33; use of drugs to dissolve and prevent formation of, 132-34

Blood transfusions, 104-105; start of blood typing, 105

Blood typing, 105

Bone marrow, storage of, 120

Bonner, James: on selective breeding, 19-20

Boulding, Kenneth E., 89-90

Brain, the, 145-71; attempts to measure effects of emotional stress on, 160, 161; cerebellum, 148; cerebral cortex, 148-50; cerebrum, 148; corpus callosum, 149; electrical activity of, 151; electrical stimulation of (ESB), 152 ff; electro-probe technique for treatment of epilepsy, 151-52; examination of the effect of external force fields upon, 158-61; invention of the electroencephalograph (EEG), 151; "memory pill," 168-69; memory-storage area of, 152-53; memory-storage mechanism, 168; pleasure center within, 153; RNA experiments in learning and memory, 164-71; stem (medulla oblongata) of, 147-48; transplants of, from one monkey to another, 115, 116 *ill.*, 118 *ill.*; see also ESB

Brain Research Institute at the University of California at Los Angeles, 158-59

Brain waves, study of, 151

Brainwashing, 145

Breeding, selective, *see* Selective breeding

Brown, Harrison, 89-90

Budding of microspheres, 13, 14 *ill.*, 15

Burnet, Macfarlane, 106

Caesarean section, 59, 60 *ill.*, 62 *ill.*

Calhoun, John B.: experiments with rat populations, 83-87

California Institute of Technology, 19

Calvin, Melvin, 16

Cameron, Ewen, 168

Cancer, 36

Cardiogenic shock, 136; methods being developed for treatment of, 136-37

Cardiovascular disease, *see* Heart disease

Carlson, Elof Axel, 39

Caton, Richard, 151

Cavendish Laboratory, 26

Cell differentiation, theory of, 48-49

Cerebellum, functions of the, 148

Cerebral cortex: functions of, 148-50; "split-brain" surgery of, 150

Cerebrum, functions of the, 148

Chemical evolution of life, theories of, 6 ff; "hot dilute soup," 9; microspheres, 11-15; polymerization, 10 ff; process of fermentation, 17

Cherry, George, 134-35

Child beating, 86

Cholesterol: role of, in heart disease, 128-29

Christian, John J., 83, 87

Chromosomal abnormalities, 67-68

Chromosomes, 24 *ill.*, 38-40; "X" and "Y," 71, 73

Cleavage of the fertilized egg, 46-47

Clonal-selection theory of antibodies and antigens, 107-108

Clotting, *see* Blood clots

"Contemporary cells," 11, 15

Contraceptives: birth-control pill, 94-95; development of a pill for men, 99 *ill.*; IUD (intra-uterine device), 95-96; long-acting birth-control capsule, 97, 97 *ill.*, 98, 98 *ill.*, 99; "morning-after pill," 93 *ill.*, 100; see also Birth control

Corneal transplants, 109, 115; Eye Bank for Sight Restoration, 118

Coronary arteries, blockage of, 128-34; blasted open with carbon dioxide under pressure, 131, 131 *ill.*, 132, 132 *ill.*; development of drug that will form soft blood clots, 133; role of the blood clot in, 131-133; role of cholesterol in, 128-29; surgery for, 129-30; use of drugs to dissolve and prevent formation of blood clots, 132-134

Coronary thrombosis, *see* Coronary arteries, blockage of
Corpus callosum, function of the, 149-150
Crick, Francis H., 26; construction of the double helix, 26
Cryobiology, 119

Darwin, Charles: on the chemical evolution of life, 5; *The Origin of Species*, 5
Death: need for a new definition of, 115-17
De Bakey, Michael, 138
Delgado, José, 153-57; animal experiments with ESB, 153-57
Deoxyribonucleic acid, *see* DNA
"Desperation" transplants of animal parts, 121
DMSO (Dimethylsulfoxide), 120
DNA (Deoxyribonucleic acid), 15, 25-44, 69; cell differentiation, theory of, 48-49; cell division and, 27-28; discovery that genetic information was transferred by, 25; electron micrograph of, 34 *ill.*; genetic control and, 30-44; man-made, 33-35; mutation and, 31 *ff*; nucleotides of DNA ladder, 26-29; operon theory, 38-39; Phi-X DNA, 34-35; polymerase, 33, 35, 36; possible uses of, 30 *ff*; synthesized by Arthur Kornberg, 33-34, 34 *ill.*, 35; thymus and, 108; triplet code of, 29-31; *see also* DNA molecule; RNA
DNA molecule, 48; coding contained in, 27 *ff*; construction of the double helix, 26; X-ray pictures of, 26
Double helix, 33; construction of, 26; description of, 26-27, 27 *ill.*; nucleotides of, 26 *ff*; *see also* DNA
Down's syndrome, 67-68

Earth's primitive atmosphere, experiments based on, 6 *ff*
Ectoderm, 48-49
Edwards, Robert, 71-73
Egg, fertilized, *see* Fertilized egg
Electrical stimulation of the brain, *see* ESB

Electroencephalograph (EEG), 151, 161, 162
Embryo: development of, 53-54; *see also* Fetus; Placenta
Endoderm, 48-49
Enzymes, 31-33
Epilepsy, electroprobe technique for treatment of, 151-52
Erythroblastosis fetalis, *see* RH disease
ESB (electrical stimulation of the brain), 152 *ff*; doubts as to future use of, on the human population, 157-58; José Delgado's experiments with, on animals, 153-57
Estrogen, 51, 93
Eugenics, 19
Evolution, theory of, 5
Evolution of life, chemical: theories of, 6 *ff*
Eye Bank for Sight Restoration, 118

Famine: predictions of, in the "Third World," 87-88
Fermentation, process of, 17; *see also* Chemical evolution of life, theories of
Fertilization, 46; sperm cell penetrating an egg, 47 *ill.*
Fertilized egg, 94; blastocyst, 72, 73; ectoderm, 48-49; endoderm, 48-49; layers of cells, 48-49; mesoderm, 48-49
Fetology, 46-77, 113; Barr bodies, 71-72; fetal surgery, 75 *ill.*, 76-77; future of, 68-77, 113; future possibility of advance determination of sex, 71 *ff*; karyotyping, 66-67, 72; research at Oregon Regional Primate Center, 69-70, 70 *ill.*, 74 *ill.*, 75 *ill.*, 76; sex-linked diseases and, 73; study of fetal chromosomes, 66-67; *see also* Amniocentisis; Artificial womb; Embryo; Fetus; Placenta; RH disease; Umbilical cord
Fetus: development of, 54-55; electronic monitoring of, 59-60; views of, in artificial womb, 58 *ill.*; *see also* Embryo
Fibrin Stabilizing Factor (FSF), 133
Follicle stimulating hormone, *see* FSH
Fortune magazine, on hunger, 87-88
Fox, Sidney, 8, 10-11, 15
Franklin, Rosalind, 26

177

Freda, Vincent, 64-65, 76
Friedman, Meyer, 129
FSH (follicle-stimulating hormone), 92-94; function of, 92

Galvani, Luigi, 151
Gamow, George, 28-29
Gardner, Richard, 71-73
Gay, Richard, 166-67
Gazzaniga, Michael, 149-50
Gellis, Sidney, 46, 77
Gene pool, 23
Genes, 23-24; Mendel's work on transmission of, 23-24
Genetic control: DNA and, 30-44
Genetic defects: detection of, in fetus, 66-67
Genetic seed bank, 21-22
Ghidoni, John, 142
Goodlin, Robert: artificial womb developed by, 56 *ill.*, 57, 58 *ill.*
Goodman, Morris, 122
Gorman, John, 65
Gurdon, J. B., 40-42; parthenogenesis experiments of, 40-42
Guttmacher, Alan, 96

Haldane, J. B. S., 9
Hamburger, Jean, 104, 105
Harada, Kaoru, 8
Harvard University: Biological Laboratories at, 8; Center for Population Studies, 80
Heart, 123-43; description and function of, 126-27, 127 *ill.*, 128
Heart, artificial: clotting, a major problem in the development of, 140-41; current work on the development of, 139, 141 *ill.*, 143 *ill.*; need for the development of, 139; and valves, 141-142
Heart attacks, 123-24, 129; cardiogenic shock, as a cause of fatal attacks, 136; drugs and devices being developed to serve as first-aid measures for, 134-37; *see also* Cardiogenic shock
Heart disease, 123 *ff*; balloon catheters used in treatment of, 136-38; blockage of coronary arteries, 128-34; development of ventricle-assist pumps, 138, 138 *ill.*, 139 *ill.*, 140 *ill.*; new drugs, dietary changes, and surgical procedures to conquer and prevent, 124-25; number of deaths due to, 123-24; *see also* Coronary arteries, blockage of; Heart attacks
Heart surgery, 129-30
Heart transplants, 102, 115, 123, 125-26, 142
Heart valves, artificial, 141-42
Heath, Robert, 157
Hess, Walter, 151
Hirose, Teruo, 130
Hitchcock, Claude R., 122
Hon, Edward H., 59-60
"Hot dilute soup," 9; *see also* Chemical evolution of life, theories of
Hufnagle, Charles, 122
Hydén, Holgar, 164-65
Hydrodynamic changes in blood-flow patterns as a cause of clotting, 141-42

IAP (Intra-Aortic Pump), 137-38
Immunological defenses of the body, 106-107, 109; antibodies *vs.* antigens, 106-107
Immuran: use of, to prevent rejection of transplants, 109
India: attitude of people toward birth control, 90
Infant mortality and birth defects, 45-46
Infanticide, 91
Institute of Molecular Evolution, 10, 14 *ill.*
Intra-uterine device, *see* IUD
Intra-uterine transfusion, 63-64; 64 *ill.*
Ireland: effect of the introduction of the potato on, 89-90
IUD (Intra-uterine device), 95-96; action of, 96; drawback of, 96

Jacob, François, 38
Japan: National Institute of Genetics, 37
Johnson, Aran, 137

Kantrowitz, Adrian, 102, 137-39
Kantrowitz, Arthur, 138, 142
Kaplitt, Martin, 131
Karyotyping, 66-67, 72
Kidneys: description and function of, 103-104; storage of, for transplants, 121; successful transplants of, 105,

110-11, 114; transplants of, 101-102, 104-105, 110-11, 114, 115
King's College (University of London), 26
Kolff, Willem, 139-40, 141 *ill.*, 143 *ill.*
Kornberg, Arthur, 32-37; DNA synthesized by, 33-34, 34 *ill.*, 35
Krech, David, 146, 165
Kuhn, Leslie A., 136-37

Learning and memory: RNA experiments with, 164-71
Lederberg, Joshua, 122
Leningrad Institute of Blood Transfusion, 118
Leukocytes, 108
Leutenizing hormone (LH): functions of, 93-94
Life, theories on creation of, 3-18; *see also* Chemical evolution of life, theories of
Liley, Margaret, 55
Liley, William, 46, 63-64
Lillehei, C. Walton, 123, 126
Lillehei, Richard, 120
Lin, Teh Ping, 43
Liver transplants, 112-13
Lorand, Laszlo, 132-33
Los Angeles Transplantation Society, 121
Lucey, Jerold, 46
Luyet, Basile J., 119
Lymphocytes, 108, 109, 111

McConnell, James, 145-46, 165-66
Magoun, H. W., 147
Markert, Clement L., 80
Mead, Margaret, 91
Medulla oblongata, 147-48; functions of, 147-48
Memory-storage mechanism, 168; *see also* Brain, the
Mendel, Gregor Johann, 23-25; formulation of laws governing genetic inheritance, 23-24
Merrill, John P., 110-11
Mesoderm, 48-49
Microspheres, 11, 11 *ill.*, 12-15; similar to Bacillus cereus, 12 *ill.*, 13; reproduction of, 13, 14 *ill.*, 15; *see also* Chemical evolution of life, theories of

Miescher, Frederick, 25
Miller, Stanley, 6-7
Mind, the: science of, 146; *see also* Brain, the
Mitosis, 28
Molecular Evolution, Institute of, 10, 14 *ill.*
Molecules, organic, 7 *ff*
Monod, Jacques, 38
Moyer, K. E., 158
Muller, Hermann J., 20-22; on changes produced by X-rays, 21; idea of a genetic seed bank, 21-22; on "improving the breed" of men, 21-22
Mutation: DNA and, 31 *ff*
Myers, M. Bert, 134-35
Myers, Ronald, 149

Nägeli, Karl Wilhelm von, 24
National Institute of Genetics in Japan, 37
National Institutes of Health, 83
Nawa, Saburo, 37-38
Nirenberg, Marshall W., 30, 31, 44
Nossal, G. J. V., 107-108
Nucleotides, 26; that make up rungs of the DNA ladder, 26-29; *see also* DNA; Double helix

Ochoa, Severo, 30
Olds, James, 152-53
Oparin, Alexander, 6
"Operon" theory, 38-39; *see also* DNA
Oregon Regional Primate Center: fetology research at, 69-70, 70 *ill.*, 74 *ill.*, 75 *ill.*, 76
Organ banks: cryobiology, 119; Los Angeles Transplantation Society, 121; need for establishment of, 118-21; obstacles encountered in storing whole organs, 119-20
Orgin of Species, The (Darwin), 5
Ortho Research Foundation, 65
Ovulation, 92

Parthenogenesis, 40-42
Pasteur, Louis, 4
Pasteur Institute in Paris, 38
Penfield, Wilder, 151-52, 153
Peptides, 10
Petrucci, Danielle, 47 *ill.*
Phenylketonuria, 32

Phi-X DNA, 34-35
Phoenix, Charles, 69-71
Photophosphorylation, 17
Photosynthesis, 17
Pickering, Donald, 56; artificial womb developed by, 57
Pirofsky, Bernard, 113-14
Placenta, 51-53, 55, 94; allowing antibodies to pass from mother to fetus, 113; amniotic sac of, 52; functions of, 51-53, 113; as an immunological filter, 113; villi of, 52; see also Embryo; Fetus; Womb, artificial
Planned Parenthood-World Population, 96
Platt, John R., 41; on genetic control, 43-44
Pollack, William, 65
Polymerase, 33, 35, 36; see also DNA
Polymerization, 10
Polyoma virus, 36
Population-control techniques, primitive, 91
Population Council at Rockefeller University, Biomedical Division of, 97
Population explosion, 79-82, 87-89
Population growth: studies of the rat, 83-87
Population growth, world, 79 *ff*; lack of leveling influences, 82; need for increased food supply, 87 *ff*; predicted effect of the proper application of technology on, 89; predictions of famine in "Third World," 87-88; "utterly dismal theorem," 89-90; see also Birth control
Population Studies, Harvard University Center for, 80
Pouchet, Felix, 4
Precambrian fossils, 16
Prisk, Bert, 137
Progesterone, 51, 55, 94, 97
Protein, 31, construction of, 6-11, 29
Protein molecules, 7 *ff*
Proteinoids, 10-11, 11 *ill.*, 13; see also Chemical evolution of life, theories of

Q-beta virus: RNA and, 36-37

Radiation: use of, to prevent rejection of transplants, 109

Ramsey, Elizabeth M., 51
Raphaelson, Alfred, 166-67
Rats, population studies of, 83-87; "battered-child syndrome," 86; "behavioral sink," 86
Redi, Francisco, 4
Reemstma, Keith, 121
Rejection of transplants, 104-105, 109, 112; methods used to prevent, 109 *ff*
Replicase, 36; see also RNA
Resko, John, 69, 71
Revelle, Roger, 80, 90
RH disease (erythroblastosis fetalis), 60-66; cause of, 60-61; development of RhoGAM vaccine to prevent, 65-66; exchange transfusion, as treatment for, 61-62; intra-uterine transfusion, as treatment for, 63-64, 64 *ill.*; mortality rate, 62-63; use of amniotic tap to detect, 63
RhoGAM vaccine, 65-66
Ribonucleic acid, see RNA
Ribosomes, 30
RNA (Ribonucleic acid), 15, 29-38; description of, 29-30; DNA molecule and, 29 *ff*; experiments in learning and memory, 164-71; Q-beta virus and, 36-37; replicase, 36; see also DNA
RNA virus: synthesized and biologically active, 36
Rockefeller Institute in New York: DNA experiments of, 25
Rosenzweig, Mark, 165

Sawyer, Philip, 131-32
Schopf, J. W., 16
Segal, Sheldon, 96-99
Selective breeding, 19 *ff*; plan of Hermann J. Muller, 21-22
Sex determination in advance: future possibility of, 71 *ff*
Sheleznyak, M. C., 93
Shemington, Francis, 145
Shettles, Landrum B., 73
Sinsheimer, Robert, 34, 35
Smith, Theobald, 65
Snow, C. P., 88-89, 100
Sobel, Sol, 131
Sperm cell: penetrating an egg, 47 *ill.*; see also Genetic seed bank
Sperry, R. W., 149-50

Spiegelman, Sol, 36-37; biologically active RNA virus synthesized by, 36
Spontaneous generation, 3-4; arguments against, 4
Starzl, Thomas, 111-12

Terasaki, Paul, 110-12
Thalidomide, 57
"Third World," predictions of famine in the, 87-88
Thymus, the, 108; DNA and, 108
Tissue-typing service, 110-11
Transplants, 101-22; corneal, 109, 115; "desperation" transplants of animal parts, 121; experiments on unborn calves to render them genetically neutral for, 122; future of, 114-22; heart, 102, 115, 123, 125-26, 142; kidney, 101-105, 110-11, 114, 115; liver, 112-13; methods used to prevent rejection of, 109 *ff*; monkeys' brains, 115, 116 *ill.*, 118 *ill.*; moral and legal problems involved, 115-17; need for establishment of an organ bank, 118-21; need for a new definition of death, 115-17; problem of storing whole organs for, 119-20; rejection of, 104-105, 109, 112; source of organs for, 114 *ff*; tissue-typing service for, 110, 111, 112; use of ALG to prevent rejection of, 111-12; use of Immuran to prevent rejection of, 109
Triplet code of DNA, 29-31
Twenty-first century, the: prospects for, 18, 19-22, 26, 31, 36, 39, 44, 46, 71, 76-77, 87, 99, 114-15, 120-22, 125-26, 134, 142, 145, 157-58, 171

Umbilical cord, 59; function of, 52
Ungar, George, 167
Urey, Harold, 6-7, 9; on the earth's primitive atmosphere, 6-7
Urinary system, 103-104; efficiency of, 104
Urokinase: use of, to prevent clotting, 132
"Utterly dismal theorem," of world population growth, 89-90

Valenti, Carlo, 66-67
Ventricle-assist pumps being developed, 138, 138 *ill.*, 139 *ill.*, 140 *ill.*
Ventricular sleeve, 135-36
Vinberg, Arthur B., 129
Vitalism, theory of, 4

Wald, Dr. George, 8-9, 17
Washkansky, Louis: recipient of first heart transplant, 123
Watson, James D., 25-26; construction of the double helix, 26
White, Robert, 115-17
Whitney, Donald, 137
Wilkins, Maurice H. F., 26
Womb, artificial, 56, 56 *ill.*, 57, 58 *ill.*, 59
World population growth, *see* Population growth, world

X-ray crystallography, 26
X-rays: changes in chromosomes produced by, 21

Yale-New Haven Medical Center, 59

Zamenhof, Stephen, 68-69

181

601

601
Warshofsky, Fred   DISCARD
The Control of Life

ORANGEVILLE PUBLIC LIBRARY.